INFORMATION, INNOVATION and IMPACTS

Economics of Science, Technology and Innovation

VOLUME 17

The titles published in this series are listed at the end of this volume.

Information, Innovation and Impacts

Edited by

JOHN DE LA MOTHE
PRIME, University of Ottawa

And

GILLES PAQUET
PRIME, University of Ottawa

Distributors for North, Central and South America:
Kluwer Academic Publishers
101 Philip Drive
Assinippi Park
Norwell, Massachusetts 02061 USA
Telephone (781) 871-6600
Fax (781) 871-6528
E-Mail <kluwer@wkap.com>

Distributors for all other countries:
Kluwer Academic Publishers Group
Distribution Centre
Post Office Box 322
3300 AH Dordrecht, THE NETHERLANDS
Telephone 31 78 6392 392
Fax 31 78 6546 474
E-Mail <services@wkap.nl>

Electronic Services <http://www.wkap.nl>

Library of Congress Cataloging-in-Publication Data
Information, innovation, and impacts / edited by John de la Mothe and Gilles Paquet.
p.cm. -- (Economics of science, technology, and innovation ; ISSN v. 17)
"This volume is one outcome of a second Statistics Canada/PRIME joint research workshop that was held in Ottawa Canada during March 1999 ..."p.ix.
Includes bibliographical references and index.
Romanized record.
ISBN 0-7923-8692-2 (alk paper)
1. Information technology--Canada--Congresses. 2. Technological innovations--Canada--Congresses.3.Organizational change--Canada--Congresses. 4. Information technology--Congresses. 5. Technological Innovations--Congresses. 6. Organizational change--Congresses. I. de la Mothe, John. II. Paquet, Gilles. III. Series.

HC120.T4 153 1999
338.9'26--dc21 99-047412

Printed on acid-free paper. Printed in the United States of America

Contents

Contributors

John de la Mothe is with the Program of Research on International Management and Economy (PRIME), Faculty of Administration, University of Ottawa.

Fred Gault is with the Science, Innovation and Electronic Information Division, Statistics Canada.

Daood Hamdani is with the Science, Innovation and Electronic Information Division, Statistics Canada.

Georg Licht is at the Zentrum für Europäische Wirtschaftsforschung (ZEW), Mannheim.

Susan McDaniel is with the Department of Sociology, University of Alberta.

Ian Miles is with the Centre for Research on Innovation and Competition (CRIC) and the Program of Policy Research in Engineering, Science and Technology (PREST), University of Manchester.

Keith Newton is with Carleton Research Unit on Innovation, Science and Engineering (CRUISE), Carleton University.

Gilles Paquet is with PRIME, Faculty of Administration and the Centre on Governance, University of Ottawa.

George Sciadas is with the Science, Innovation and Electronic Information Division, Statistics Canada.

Nico Stehr is at the Max-Planck-Institut für Meteorologie, Hamburg.

Valerie Steeves is with the Department of Law, Carleton University.

Graham Vickery is in the Division for Science, Technology and Industry at the Organization for Economic Cooperation and Development (OECD), Paris.

Laura Winer is with the Network of Centers of Excellence on Telelearning, and is based at the University of Montreal.

Cliff Wymbs is with Baruch College, City University of New York.

Acknowledgements

This volume is one outcome of a second Statistics Canada / PRIME joint research workshop that was held in Ottawa Canada during March 1999 on the general theme of *Information, Innovation and Impacts*. The first one had been held in March 1997 and resulted in the 1998 book, also published by Kluwer in this series, that we co-edited on *Local and Regional Systems of Innovation*.

The focus of this second workshop was information – principally information that was processed and transferred electronically – on the innovative use of this information, and on its social, organizational and industrial impacts. The purpose of the workshop was to help set the research agendas for statistical measurement, academic research and policy development.

Much of the social and economic changes induced by innovative uses of information lend themselves to measurement but this cannot be done well without rich underlying conceptual frameworks and understanding that provide a basis for the development of sound measurements. So it is crucially important to develop analytical categories that map both the components of the innovations in information processes and products, and the impacts they have had on people, on the ways in which they organize and do business, on the ways concerns organize to take full advantage of these innovations and to effect faster learning.

These impacts raise important policy questions. What about the differential accessibility to the new informational infrastructures or the differential capabilities to make use of the access that is provided? These two issues – access and capability – are examined with special reference to certain economic realities like electronic commerce as there is already much work done at the international level in these areas. But the scope of the workshop and the book is much wider than e-com. What we were seeking was a broad view of the socio-economic insights into the new innovative uses of information and information and communications technology.

In bringing this project together, from the idea and the workshop to this book, we would very much like to extend our thanks to: Fred Gault of Statistics Canada for conceiving of the workshop in the first place, for supporting the project and for collaborating with us in its development; several students from PRIME for their help in making the meeting a reality, particularly Tyler Chamberlin and Gyneya Dicks; colleagues in the loose Ottawa-based network under the auspices of PRIME's research program on the management of innovation systems, sponsored by the SSHRC-NRC Innovation System Research Networks; Emer Killean who quickly turned a rough elastic-bound set of papers into a camera ready manuscript; Ranak Jasani of Kluwer, for once again guiding and encouraging this project so kindly; Bo Carlson and Christiano Antonelli for inviting this book into their fine series; and each of the contributors – both those who were at the workshop and those who are represented in this volume – for their very stimulating contributions.

JdlM/GP

Preface

This book - a joint project between the Program of Research in International Management and Economy (PRIME) at the University of Ottawa and Statistics Canada - brings together economic, social and statistical views of the impacts of the innovative use of information. It examines costs as well as benefits of rapidly expanding availability of data, of information, and of codified knowledge, and it provides suggestions for future work.

This project fits into an on-going research program at Statistics Canada to develop indicators for science and technology (S&T) in a coherent manner, so that the story can be told of the activities of the actors in S&T systems, their interactions, and the outcomes as the actors go about the generation, transmission and use of S&T knowledge. A key part of this story is the transmission, or mediation, of knowledge, of information, and of data, as part of the effective functioning of the system.

Information and communications technologies (ICTs) provide the goods and services through which the information is moved, processed, stored and displayed. They also provide the means to enable all manner of teleservices, including telelearning, telemedicine and electronic commerce. All of these activities have the capacity to change the ways in which we learn, do business, and enjoy ourselves and these new ways of doing old things have impact, on people, regions, communities and businesses. Measuring those impacts, be they good or bad, is a necessary part of the feed back loop to people, governments and corporate decision- makers, if they are to have the information to change the system of which they are a part.

Behavioral change, corporate, community and individual, is an outcome of decision making by people who control resources and there is a need to make visible such change, and its causes. This is a role for statistical measurement.

Measurement is then an essential aspect of understanding what is going on in a system and that is why considerable time in this project has been devoted to examining measures of the use of ICT products by firms and people. The introduction of new or significantly improved products or processes to the market is innovation and it is the impact of the resulting Schumpeterian creative destruction that transforms economies and the lives of people. Innovative use of information and knowledge-based products is central to this transformation.

This study comes at a time when the Organization for Economic Co-operation and Development (OECD) is implementing an industry-based definition of the ICT sector and is moving on to examine measures of the 'content' products. 'Content' products are the texts, images and sounds that are stored on electronic media, like CDs, transmitted by wireline or wireless means, or are available interactively through electronic networks. This is all part of developing indicators for the Information Society so that the change in use of related technologies can be monitored, along with their social and economic impacts.

However, it is not just technologies that are changing but also the people in the society and the networks through which they interact. People are rising to progressively higher levels of academic qualification and they enter the labour force with greater expectations than they once did in respect of their capacity to act, to use the knowledge that they have acquired, and to exercise autonomy. This has impact, just as the use of ICTs and of new knowledge-based products do. The process of learning, itself, is a subject for research and is not limited to the learning person, but includes the learning firm and the learning region.

With all of these threads there is a need to set a research agenda and to move towards a framework for understanding the system. This was the motivation for a research workshop convened by the University of Ottawa in March of 1999. Some of the papers in this book are derived from that workshop, while others have been invited to complement the collection.

- The themes of the inquiry are set out in the introduction and are developed in a section of on Challenges. Broadly, they are:
- The interaction necessary for the transfer of knowledge, and the consequence of the intangibility of the knowledge product, especially in service industries;
- How organizational learning is to be achieved, how knowledge can be produced in an organization and the role of the mediator/facilitator;
- The organizational forms that are based on innovation, high skills, organizational flexibility and trust; and,

- A map of how social impacts of these activities might be captured and examined.

The section on Measurement presents the results of statistical work and analysis and, from this, it is clear that there is infrastructure in place to support the measurement of the use of ICTs, and of innovation, and that the analysis provides insights into the consequences of such use. While this is encouraging, there is still work to be done and this theme recurs in the final section.

However, not all use of ICTs is necessarily beneficial and not all people can access the new wonders of the information society, and these issues are raised in the section on Impacts and are brought together in a policy analysis. The key themes are:

- The impact of electronic commerce on products and their delivery, on foreign direct investment, and on the forming of strategic alliances;
- The challenge of networked communications to existing legal concepts of privacy and property;
- The implications of globally available telelearning for culture, life-long learning, and the blurring role of teacher and student, leading to questions of physical, functional and equitable access; and,
- The thesis that the 'productivity paradox' is driven by a transformation of society and of the work of work, rather than being an artifact of technological change

A recurring theme in the papers is trust, in networked interactions and across space, as part of the innovative process. The spatial aspect of innovation was the subject of an earlier volume in this Kluwer series (*Local and Regional Systems of Innovation*, also edited by John de la Mothe and Gilles Paquet and published as Volume 14 in the Series 'Economics of Science, Technology and Innovation') and is consequently not developed here.

The concluding chapter tells us where we are in understanding a complex picture, but this is not the end, it is a taking of stock along the way, and a suggestion for next steps in our journey

Fred Gault

Director, Science, Innovation and Electronic Information Division,
Statistics Canada

PART I

CHALLENGES

Chapter 1

Informational Innovations and Their Impacts

JOHN DE LA MOTHE AND GILLES PAQUET[1]
PRIME, University of Ottawa

"To understand the world and act on it, we need workable concepts."
Geoff Mulgan

1. INTRODUCTION

While the old economy was driven by manufacturing and tangible commodities, and was focused on the allocation of existing material resources, the new economy is knowledge-based and technology-driven. It is geared towards innovation, and to the creation of new use-values, products and services.

The new competition is distinguished from the old by the fact that, in the old days, the firm took its milieu as given (technologies, institutions, preferences). The old competition tried to optimize within this context by choosing the "right" technology and the "right" product mix. It competed ferociously through prices with other firms, and counted on little help from consumers, suppliers or governments. The new competition is generated by globalization, integration, and accelerated technical change. This is forcing enterprises and institutions in the new economy to embody a philosophy of continuous improvement and innovation, to become learning organizations, and to do so, they require an organizational flexibility that they did not previously possess. This includes adopting a consultative coordination capability within the production and value chains, and a cooperation strategy with other stakeholders and governments.

Producing and distributing new knowledge has become centrally important, and firms have had to organize for knowledge. Their value-creating capacity depends on it. (Best 1990; Wikstrom and Normann 1994) Moreover, since, in the dynamic, innovative and cooperative environment of the learning economy, the capacity to learn increasingly determines the relative position of individuals, firms and national systems, new modes of production of knowledge and new modes of collegiality, alliances and sharing of knowledge have evolved (Gibbons *et al.* 1994; Lundvall and Johnson 1994).

But such processes do not necessarily work perfectly. While much know-what and know-why has been ever more effectively codified, and can be produced and distributed as quasi-commodity, know-how and know-who have remained tacit and socially embedded (Foray and Lundvall 1996). Consequently, the production and the distribution of this latter form of knowledge has been more problematic: it depends a great deal on social cohesion and trust, on much trespassing and cross-fertilization between disciplinary fields, and on the development of networks capable of serving as two-way communication links between tacit and codified, private and shared knowledge, between passive efficiency-achieving and creative/destructive Schumpeterian learning (Boisot 1995). There are ample possibilities for coordination failures that can down the process of learning.

In this information-rich context, communication is obviously crucially important. And the arrival on the scene of new information and communication technology (ICT), with its capacity to transfer information immensely faster and to connect an ever larger number of partners, has accelerated even more the pace of change. In a world of ever greater competition, these informational innovations have triggered a dramatic series of shockwaves that have accelerated the innovation process per se but also transformed the ways in which the economy is structured and functions at the micro, meso and macro levels.

The purpose of this chapter is to provide a general conceptual framework to help us analyze the impacts of these new synergies between information and innovation on modern socio-economies at the micro, meso and macro levels.

We proceed in five stages. First we sketch the broad context within which these developments have been nested: a new context where the economy is innovation-driven and has become truly a learning socio-economy. Second, we collectively probe the central features of the new socio-economy: the emergence of a new cognitive division of labour with all the implications it is bound to have on the decision-making and production process, but also on the organizational fabric of the socio-economy. Third, we identify four families of challenges that have been generated by these new information-innovation synergies, and we show how they have shaken a variety of certitudes that used to be in good currency within the social-scientific and management literature. Fourth, we develop a rough analytical framework to help us in sorting out the impacts of these informational innovations. This framework identifies four families of mediating agencies between the innovation system and the socio-economic process, and probes the dynamics of mediation between them but also with the socio-cultural context within which they are both embedded. Fifth, we underline some features of the chapters in the rest of this book and show to what extent the authors have thrown some light on this dynamics of mediation. It should be clear that in the case of many issues, a complete assessment will have to await a longer period of experimentation.

The general objective of this project is to analyze the socioeconomic impacts of these information-innovation synergies in order to contribute to the on-going process by which Statistics Canada continually refines the sort of metrics it uses to gauge socio-economic change in the new economy.

Some of the challenges faced by Statistics Canada and other national statistical agencies call for simple enough adjustments to current data collection procedures. Others will require different survey instruments to determine patterns of use of the new technology, and diffusion processes, but also the impacts on organizational forms, employment patterns, lifestyles and socio-cultural impacts of the direct and indirect proliferation of the new ICT. In all of this, as we will see, research and data collection on attitudes is bound to become much more important.

An indication of this is recent efforts at the OECD leading to an industry-based definition of the ICT sector, and its on-going efforts to define the products produced by the sector and used there and throughout the economy. Work at the OECD is now turning to the classification of the content that is displayed, processed, stored and transmitted by ICT goods and services. All of this is essential to building a realizable research and measurement agenda.

In the last part of this volume, we go beyond this framework to help design a refurbished set of statistics capable of apprehending the new economy more effectively, and introduce broader considerations to be kept in mind in designing public policies to deal with the impacts of the new information technologies.

2. THE LEARNING ECONOMY[2]

Neoclassical economics has always had difficulty dealing with information unless it could be regarded as a commodity. Markets handle commodities well, but they do not process non-commodities as effectively. By reducing information to information bits, neo-classical economists thought that they had domesticated information, but, in fact, the problem had simply been defined away (Paquet 1994). It became clear in the 1980s that knowledge, communication, and learning could not be reduced to information bits.

Dissatisfaction with this approach led to a shift in the language used to describe the new economy in the late 1980s. At that time, the expression "knowledge-based economy" acquired good currency. It acknowledged the fundamental difference between ideas and commodities, and displaced the notion that information or knowledge could be regarded as "quasi-commodities". "Information" came to connote codified and embodied propositions or algorithms, while "knowledge" embodied a much broader range of realities — codified knowledge, but also uncodified or tacit codes, skills, heuristics, etc. (Dosi 1996).

Instead of focusing on information bits, economists began to build on the fact that the economy was now rooted most strongly in the production, distribution and use of knowledge. There was still a certain degree of commodification. This is most certainly the case for codified knowledge which was regarded as a quasi-commodity. The knowledge-based economy has, in much the same manner as the information economy, come to be X-rayed mainly through a loose categorization of activities and employment (Machlup 1980–84).

This shift from information to knowledge has increasingly displaced attention toward the process of learning, and this may have been its most important impact. It has led, in the last five years, to an extraordinary explosion of work centred on learning and cognition (Malerba 1992; Lundvall and Johnson 1994; OECD 1996).

This latter shift toward a *learning economy framework* is not one of degree, but is, instead, one of kind. The new economy, under this label, is defined as innovation-mediated and innovation-driven. The learning capabilities of individuals and organizations are the source of wealth creation, but most importantly, the learning economy is rooted in a social or collective mobilization of knowledge. Learning is about harnessing the collective intelligence of the team as a source of continuous improvement (Florida and Kenney 1993). This in turn commands a degree of cooperation to take advantage of positive externalities, increasing returns, and strong cumulative experience learning processes (Jacquemin 1995).

The learning economy is therefore built on two pillars: cognition and interaction.

Cognition refers to the process of extraction of information from the environment through perception, and the development of knowledge through communication. In this framework, information exists to the extent that it becomes embodied in the brain as pattern or in an artifact (capital) that is designed to act as a surrogate for the brain. Instead of focusing on the allocation of existing informational resources, cognitive economics focuses on the production of new knowledge.

To the extent that cognition is conditioned and restricted by the mechanism of brain and mind, it is all but impossible to understand how information and communication matter without a fair idea of the way in which people gain knowledge. This is the main reason for the focus of cognitive economics on the protocols leading to cognition.

The brain registers sensory inputs. These inputs are information extracted by perceptual systems. In the beginning is the percept. The perceptual systems search the environment for behaviourally relevant change and extract information patterns from the invariants in the environment. This information is then registered in the brain in the form of neural nets (i.e., of interactive assemblages of neurons in patterns). A new experience creates a pattern recognition cascade often built from existing patterns, plus a small set of new patterns generated as new neurons fire in response to input. A non-trivial message therefore triggers a reconfiguration of existing patterns through the firing of new neurons (Paquet 1989; Coward 1990).

Cognition is the source of meaningful (behaviourally-relevant) information, and cognitive economics aims at developing a higher ratio of meaningful information-to-noise through an improvement of the various communication enhancement mechanisms: more effective skills at extracting patterns, easier transformation of frames of reference, reduction of cognitive dissonance, and so on.

But cognition does not occur only at the individual level. Creative individuals can rarely reach their goals in isolation. They operate in the context of organizations embodying a communication network. Organizations learn from experiences in real-world groups (Hutchins 1991). When the informational environment changes, as a result of new circumstances or breakdowns/losses of perceptual equipment, new stable work configurations are generated that are not unlike modified neural networks.

Interaction, the second pillar, is necessary to generate effective learning. It focuses on the desirable form of imperfect competition, or mixes of competition and cooperation that characterizes the learning economy. Learning entails "the mutually consistent interpretation of information that is not fully codified, and hence not fully capable of being transmitted, understood, and utilized independently of the actual agents who are developing and using it" (Storper 1996:259).

Interaction, therefore, is of central importance because of the fact that knowledge is dispersed and exists in a form that is not fully codified. This calls for conventions or relational transactions to define mutually coherent expectations and common guideposts. These conventions differ from industry to industry, but they play somewhat similar roles in each sector: they provide the requisite amount of coherence for a common context of interpretation, and for some "cognitive routinization of relations between firms, their environments, and employees" (Storper 1996:259).

These conventions are the social ligatures that generate the coherence of this network economy. Yet one of the paradoxical characteristics of the network economy, and a major source of its nimbleness — of its capacity to learn — is the fact that it actively seeks and embodies creative sustainable disequilibrium. A good learning network must not be too coherent: it should connect the different nodes enough that they can learn from one another, but the nodes should not be too similar, nor the ties too strong or too routinized if creative tension is to exist, and if learning is to occur from one to the other. This is the sense in which one may speak of "the strength of weak ties" (Granovetter 1973): a certain degree of heterogeneity, and therefore social distance, might foster a higher potentiality for innovation because the different parties bring to the "conversation", not replicate knowledge, but a more complementary body of knowledge. More fruitful synergies ensue.

The new economy is thus fundamentally a learning economy that tries to connect everything to everything else, in an ever expanding network. The laws of this new cognitive and coordinated economy are perplexing, to say the least, for those still in the mind set of the industrial age (Mulgan 1997; Paquet 1998).

First, there is a dramatic increase in the ratio of the exploration of new avenues to the exploitation of existing resources, and this challenges much of what was the conventional wisdom of the old industrial age (March 1991).

Second, and more importantly, in this new counter-intuitive economy, connection pays, plentitude creates value exponentially as marginal costs shrink, irreversible momentum is achieved at low threshold, increasing returns are everywhere, items become cheaper as they improve, it pays to give your product free of charge, devolution pays, perpetual disruption is the only sustainable way of life, and effectiveness is more important than efficiency (Kelly 1998).

3. TOWARD A COGNITIVE DIVISION OF LABOUR

The transition to the learning economy has meant a drift toward a new division of labour, but there is some disagreement about the exact nature of this evolution. Some see it as a change strictly at the organizational level: the emergence of networks of small flexibly specialized firms or the return to the dominance of big firms. Others have a richer perspective, focused on the fabric of the emerging new shop floor as technological and organizational changes combine to "harness the collective intelligence of workers as a source of continuous improvement" (Florida 1991:560-1). On the new shop floor, "knowledge is in effect socially and/or collectively created. Thus, the capabilities or "brains" of a variety of different types of workers are integrated and explicitly harnessed in the process of turning knowledge into commodities and new productive forces".

Since knowledge is traditionally *segregated* into institutionally imposed sub-divisions, and much of it is *tacit* (i.e., heuristic, subjective, embodied knowledge not easy to articulate) and *dispersed* widely amongst agents, what are required are new techno-organizational forms that are likely to aggregate, tap into and harness all these capabilities (Minkler 1993; Senker 1995). This is essential if learning is to proceed. These new ways of creating value are forcing an integration of mental and manual labour into *self-managing work teams* that bear little resemblance to the traditional division of labour: "Teams are used to develop links to and connections across the innovation-production spectrum" (Florida 1991:571). We have moved beyond the Taylorian system where this spectrum was split into self-contained and isolated segments which shared little knowledge. The self-managing work team grows in all directions in order to internalize, through all forms of *relational contracting*, the required members needed for the team to learn effectively.

The old division of labour was a technical division between more or less generic and complementary factors of production combined in technically efficient ways. The production function describes this old reality, and it is still a staple product in most microeconomics textbooks. The focus of the factor-of-production firm is technical and allocational efficiency. This search for efficiency leads to a hyper-fragmentation of the process of production into specialized bits of activity requiring a minimal amount of skill. Indeed, the "deskilling" of production workers has been an efficient way to quickly produce large runs of homogeneous output at low cost in the typical Taylorian production process, where "*le travail est en miettes*".

In the new learning economy, continuous improvement and innovation are the driving forces. Self-managing work teams constitute cognitive entities: i.e., clusters of activities that are organized around capabilities, intellectual assets, and knowledge processes in such a way as to maximize the ability of the organization to learn (Stewart 1997). Moreover, the organization as a value-creating partner system, organized for knowledge production, is much like an organism: "Because of the tacit nature of knowledge, a firm is capable of doing things that it cannot easily describe" (Minkler 1993:583).

Obviously these different blocks of knowledge must be technically compatible, but what is most important is their capacity to "progress in concert" (Moati and Mouhoud 1994:59). In such a context, hyper-specialization generates rigidities and proves dysfunctional. The mastery of a block of knowledge requires *learning communities* or *communities of practice* that are bound by conventions, and are unlikely to be hyper-fragmented into monads. What is left at the interface between these cognitive blocks is the locus of what is most easily standardized and routinized, where learning and fast adaptation is least difficult. What is inside the blocks of knowledge is what requires finely tuned on-going learning.

Not all industries will be affected in a similar way by this shift from a technical to a cognitive division of labour. Many sectors, characterized by rather unsophisticated knowledge bases, may still function relatively well with a technical division of labour. But in sophisticated knowledge-based industries, like the modern automobile or aeronautics sectors, the shift to the new division of labour has been nothing short of epoch-making. The production process has been partitioned in dramatically different ways and reconstructed along cognitive entities lines.

This has allowed experimentation with new organization forms where sub-contractors with particular core competencies, or at the centre of important techno-organizational networks, have become responsible for substantial segments of complex products. In some cases, the parent firm is now accounting for only a small fraction of the workforce involved. In such a context, the assemblage has become apparently so relatively trivialized that it is now possible to simultaneously process different types of vehicles at the same time on the same production line. This is possible as a result of the bundling of complexity in blocks of knowledge that are responsible for substantial segments of the overall product.

In this new cognitive division of labour, harnessing dispersed, tacit, and ever changing knowledge is no simple task. Moreover, it cannot be easily done through the market. Many pursue the dream that all knowledge could be codified, and become a quasi-commodity that could be handled by the market. In this scenario, learning might become another sort of transaction in codified knowledge. This is utopian: tacit knowledge creates difficulties and bottlenecks, and incomplete knowledge is inescapable. It may even be argued that excessive codification may well be a source of inertia and deceleration in the process of learning and change (Foray and Lundvall 1997). So tacit knowledge is inescapable, and knowledge production and dissemination cannot rely simply on market mechanisms. Learning is organization-based. The organization is a locus of creative interactivity. Social or collective learning must be interactive. Simple coordination will not suffice.

Mechanical coordination may have been sufficient in the technical division of labour world, but when dealing with blocks of knowledge evolving through time, a more dynamic form of interactivity is required, to provide a form of much more complex harmonization.

Interactivity is a complex form of dialectical relations between agents and segments, and their evolution through time. It harmonizes the different capabilities or competencies of organizations (be they technical, organizational, strategic, or learning), the particular capacities of the different organizations to solve problems, to absorb knowledge, to innovate and experiment, and to incorporate new knowledge in its functions, the interactions with the environment and with other organizations, and the various rhythms of dynamic increasing returns of the different organizations through learning by learning.

Interactivity brings some sort of cumulative process of learning built on externalities of all sorts, with great potential for irreversibility and inflexibilities (Le Bas 1993:13). But it mostly provokes the genesis of institutions: a set of guideposts, the locus for the memorization and transmission of routines and tacit knowledge, through conventions, contracts and contraptions that form a cognitive framework that guides the learning process (Llerena 1997).

In the new cognitive division of labour, interactivity is the bond that enhances social learning, but since institutions organize the interactivity, they also constrain the nature of the exploration for, and the exploitation of, new knowledge. They orient the directions of learning.

4. FOUR MAJOR CHALLENGES

The learning economy has not only generated a new cognitive division of labour, but it has also brought forth a new mode of production of knowledge. While Mode 1 production of knowledge was disciplinary, governed by the norms of the scientific method, carried out in stable hierarchical academic institutions, and mainly accountable to the canons of the Republic of Science, Mode 2 production of knowledge is transdisciplinary, heterogeneous, carried out in transient heterarchical contexts of applications, and more socially distributed and therefore socially accountable (Gibbons 1994; Gibbons *et al.* 1994).

While this new mode of production of knowledge is better adapted to a world where knowledge is *de facto* segregated, tacit, and distributed, it poses immense problems of coordination as the number of stakeholders, the number of perspectives, the numerous types of accountabilities, and the required trespassings are multiplied.

It means that the traditional distinction between science and professional practice has all but disappeared, that collaborative ventures are now the rule, that co-evolution by all those involved in this production ensues in the negotiating arena where the new knowledge is produced and disseminated. Brokering this complex form of arrangements, at a time when the context of application is evolving rapidly, and when the continuous negotiation and interaction among the interested actors raises complex questions of 360-degree accountabilities, is quite a challenge. Much of the necessary co-evolutionary effects may be prevented by institutional rigidities, or by the lack of the necessary framework rules in the existing institutional order.

This reconfiguration of the technology of production of knowledge is obviously generating important challenges, for they often require a profound modification of the social technology. For Mode 2 production of knowledge, in the beginning is the issue. The context of application and the nature of the problem generate a creative process. This is the world of Delta knowledge, knowledge that is the result of reflection-in-action, not the application of already extant knowledge, but the development of learning by doing (Schon 1983; Gilles et Paquet 1991). In that sense, the framework is continually evolving in a way that is akin to the situation of the designer who tries to find some goodness of fit between two intangibles — a form that still does not exist and an evolving context that cannot be fully described since it is evolving constantly (Alexander 1964).

It is impossible to examine in this short introductory chapter the dysfunctional character of a whole range of existing institutions in the face of the new production of knowledge. But a few families of problems appear to be important enough to warrant some attention.

These problems (having to do with the lack of an adequate apparatus to conceptualize and measure these new relevant realities, with property rights, organization, ethics, public policy, epistemology, etc.) are especially important since they represent challenges that are either entirely unmet or very poorly dealt with at this time. Unless we deal with these questions effectively in the near future, the new learning economy will be significantly hampered.

First, to explore the structure and functioning of innovation systems and their impacts in the context of the new information economy poses a lexicon challenge. For we do not have a completely satisfactory picture of either the innovation process or of the *filière* within which it operates, or of the diverse impacts it has on the broader social context in our connected information society.

We use the concept of *filière* to underline the multi-dimensional and the counter-intuitive nature of the process. As *The Economist's* recent Survey reminded us, "new ideas travel in roundabout ways" and "innovators break all the rules" (1999). This does not mean that innovation processes cannot be modelled, but it will require more sophistication than has been available up to now. As a starting point, we have used the connectionist model of innovation inspired by the neural net model of John Ziman (1991). This model focuses on the cognitive dimension of the innovation process, on the capacity of the net of ideas, techniques and commodities to produce new knowledge in the form of a new pattern or a transformation of existing patterns. To understand the process of innovation in this context, one requires an approach in terms of evolutionary cognitive economics (Paquet 1998).

Moreover, the way in which the innovation system is embedded within the broader socio-economic system is not well understood. So it is difficult to track down the impacts of innovations in the information age because of the lack of a useful map of the main features of the new socio-economy that are likely to be subjected to Schumpeterian creative destruction. There is even a lack of the requisite lexicon necessary to tackle these issues. So much has to be done to construct even a provisional conceptual framework to analyse the innovation systems and their impacts.

This major flaw at the conceptual level cannot be repaired in a mechanical way without a full understanding of the new challenges brought forth by the new learning economy. These constitute a second major challenge – the structural capital challenge.

Knowledge has always been an important economic input, but new knowledge production has become so central in the new economy that new categories and reference points have become necessary. For innovation is fundamentally linked to the production of new knowledge and its impacts in a learning economy are important especially as it affects the learning abilities and the capacity to transform of the socio-economy.

Social scientists (except for psychologists) have done little work on this front until recently, but it has become a central focus of many new research programs (Guilhon et al. 1997). These programs have emphasized the dual dimensions of cognition and interaction as central to the process of innovation and social learning. In the learning economy, the exploration for, and the exploitation of new knowledge are the driving forces. This fact requires organizational arrangements, at the production level as well as at the corporate level, that are designed to ensure learning as effectively as possible. But these problems have not been studied with sufficient care to provide ready answers to the question.

What sort of organizational design is likely to resolve the coordination problem when different expectations exist, and when bounded rationality, uncertainty and asset specificity render a market solution too costly? To tackle this question, we need to understand more fully the notion of structural capital, and the ways in which structural capital may be designed to better manage tacit knowledge, and better mobilize those persons involved in the learning process (Saint-Onge 1996).

Structural capital is defined as the capabilities of the organization, and capabilities are "forms of knowledge about how to carry out productive tasks" (Langlois and Robertson 1995:16). The central point in Saint-Onge's argument is that much of this knowledge is tacit. For Saint-Onge, structural capital consists of four elements — *systems* (the way its processes and outputs proceed), *structure* (the arrangement of responsibilities and accountabilities between the members of the organization), *strategy* (the goals of the organization and the ways it seeks to achieve them), and *culture* (the sum of individual opinions, shared mindsets, values and norms within the organization) (Saint-Onge 1996:13). And the major barrier to success is the lack of fit or alignment between these four elements.

The design or redesign of structural capital amounts to the creation of new capabilities. A new organizational feature arises as an answer to a problem of coordination, a problem of alignment between systems, structure, strategy and culture. While much work has been done on organizational design, such work does not suggest the conditions necessary for the required alignment referred to above. One is always tempted to use one of the four components as a lever and to simply adjust the other three to the ruling one. This has proved rather disastrous as a search for value-creation and added value. For instance, the reengineering of systems has been presented as a panacea and has failed miserably (Ferrand and Paquet 1994). In the same manner, restructuring may capture a portion of the necessary adjustment, but is unlikely to be sufficient.

Saint-Onge suggests that, even though tacit knowledge has an impact on all four components of structural capital, culture tends to be the most implicit, and the connections between strategy and culture are likely to be the most potent. But, it is the goodness-of-fit between the four components (systems, structure, strategy, culture) that is central. The alignment and congruence of the tacit knowledge captured by human capital (assumptions, values and beliefs characterizing the mindset of individuals), customer capital (the mindset of customers that shape their perception of valued-added) and structural capital (the collective mindset of the members of the organization that shape the culture of the organization — i.e., its norms and values) also act as filters, and form a mental grid that shapes systems, structure and strategy (Saint-Onge 1996).

This nexus of forces at the core of the structural capital almost naturally breeds misalignment. Whether one tackles the production or the organizational side of the firm, the problem of lack of goodness of fit among these four elements of structural capital (systems, structure, strategy, culture), in much the same way as the lack of goodness-of-fit between the three components of intellectual capital (human, customer, structural), stands at the source of much of the firm's poor performance. These coordination failures are sometimes misunderstood, but more often than not they are simply denied or suppressed in the public discourse concerning firms. The same may be said to hold at the meso and macro economic levels.

Third, there is the evolutionary dynamics challenge. The age of connexity has transformed everything from our organizations and communities, our governance regimes, and our mechanisms of coordination to the very ecologies of our mind, our identity and our values. We cannot assume that the innovation *filière* is not also being transformed. And it would be unwise to conceptualize the *filière* or to attempt to measure it as if it were not evolving.

This is a world in which the whole governance regime is changing. There is a distributed governance, re-defining the respective role of the private, public and civic sectors, but also a degree of self-organization that has increased substantially. Indeed, in this world of interdependence, connexity means a world without a master, a broad socio-political-economic context of living organizations that cannot ever be regarded as a placid and inert environment. Innovation processes are embedded in and impact on an evolving context that, in turn, has an evolving influence on the very nature of our innovativeness and creativity.

This new dynamic goes well beyond the simple networks of exchange or the simple technological trajectories. The new learning economy is first and foremost a cognitive reality, the new division of labour is effected along cognitive lines, and the new dynamics commands that the economy organizes for learning. We are very far from the diktats of allocational efficiency. Yet these Schumpeterian dynamics have been used much more as a metaphor than as an explicit analytical mechanism.

The evolutionary cognitive paradigm, based on individual and organizational learning in a world of bounded and selective rationality, pattern thinking, pattern causality, self-organization and evolutionary change, opens new vistas and forces the revision of many prescriptions. Chief among the new terrains open to exploration by the new paradigm are three areas pregnant with new insights into the structure and functioning of modern economies (Paquet 1998).

First, one must single out the world of inert areas, selective rationality, cognitive filters and cognitive failures, that has been probed and analyzed in a mildly heterodox literature over the last decades (Leibenstein 1976; Kahneman and Tversky 1979; Akerlof and Dickens 1982; McCain 1992, Baumard 1996). Their surprising results have been interpreted as a string of anomalies. Consequently, they were either ignored or rationalized away.

These results invite an analysis of the power of mediating cognitive structures (Elster 1989; Frey 1990) and an inquiry into areas like the cognitive basis of entrepreneurship, the search for novelty, creativity, and impulse-filtering in organizations. In all those areas that have been somewhat neglected by neo-classical economics, the evolution of belief systems and cognitive structures has become a subject of more attentive analyses. They have also become the focus of public policy in a way that had been anticipated by Simon and Leibenstein: cognitive economics has simply been intent on extending and radicalizing many of their ideas on cognition.

Second, the whole area of norms, conventions, routine patterns and other organizational features (the organizational equivalent of cognitive structures) has also become of central importance. The *pattern of relationships* is the new unit of analysis: this is the world of rationales, techno-organizational patterns, networks and moral contracts. They are embodied in neural-network-type relationships that are very similar to those that embody knowledge in the brain: learning or the production of new knowledge takes the form of a new pattern of relationships, and social learning means embodiment in a new set of relationships (North 1996).

Third, the new centrality of patterns of relationships cannot but bring to the center of the stage the challenge of understanding the dynamics of evolutionary self-organized economic systems.

These issues have already caught the attention of even those most determined to work within the neo-classical paradigm (Schelling 1978; Leydesdorff and Van Den Besselaar 1994; Krugman 1996). While this new approach in terms of self-organization is nothing but another name for network learning, it suggests a new interest in network externalities, Myrdalian cumulative processes, dynamic increasing returns, cumulative technologies, and emergent properties of complex systems. It provides for interesting ways to take into consideration biased representations and cognitive structures as major elements in the shaping of path-dependent long-run trends (North 1996).

Fourth, there is a metrics and policy challenge. Not all aspects of these processes of innovations and of the impacts they have on the socio-economy are quantifiable in a meaningful way. And even those aspects that are quantifiable require much ingenuity to capture even in a partial or oblique way some of the socio-economic impacts of great consequence.

There has been a multiplicity of efforts to measure innovation in all sorts of ways. Until the 1990s, most were based on linear models of innovation in which stylized inputs like R&D spending, or stylized outcomes like patents were linked by a mysterious black box or by an opaque process labelled "ingenuity". This was hardly satisfactory. In any realistic model of innovation, there is an inherent complexity that needs to be unpacked for measurement purposes. There are important linkages and spillovers that need to be taken into account if a reasonable measure of the variety of impacts of innovation is to ensue.

This was part of the motivation for the first revision by the European Union and the OECD of the Oslo Manual (OECD 1997), which provides guidelines on the measurement of innovation. The measurement issues have been developed in the series of Community Innovation Surveys (CIS) conducted by Eurostat, the statistical office of the EU, and in surveys run by other statistical offices. These issues are discussed in contributions to this volume (see Daood Hamdani and Georg Licht). While it is difficult to measure and analyze a complex activity like innovation, along with its linkages and its outcomes, the real issues of interest are the economic and social impacts of the activity, an area of current research at Statistics Canada.

The measurement problems are even more awesome when it comes to gauging the impacts of innovations in an information society. So many forces are at work in shaping the evolution of a socio-economy that it is very difficult to ascribe to specific innovations their appropriate impacts.

This has led to excessive claims tending either to deny any impact for an innovation or to ascribe all the meaningful societal changes to it. While this makes for interesting headlines and media-hype, it cannot suffice when one is interested in developing a meaningful basis for public policy.

The new world of policy generated by the transformation of both the innovation process and the information society is ill defined. Intervening in the process of production of new knowledge is immensely complex, but it is not as hopelessly difficult as some would suggest. What is required however is a capacity for lateral thinking, creative induction, transversal conversations that are at the core of new discoveries.

5. A FRAMEWORK TO STUDY THE IMPACTS OF INFORMATIONAL INNOVATIONS

A major difficulty in dealing with structural discontinuities and robust processes of structuration is that they cannot be modelled with deterministic mathematical representations. Self-organization of the sort we observe in history involves endogeneous, non-equilibrium processes. But there is now a significant literature in the natural sciences that examines "transitional properties in the presence of self-organization", this is known as the "science of synergetics" (Foster and Wild 1996).

We are still very far from being able to count on workable economic strategies ("an economics that can deal with transition") but we may at least focus on the dimensions suggested by synergetics to construct a qualitative framework to study the impacts of innovations in an information economy. This entails an effort to identify the variables likely to be more stable or to focus on a few unstable variables at critical points.

This is more easily said than done. In open, non-stationary social systems, as many neo-Schumpeterian studies have shown, "in the presence of technical innovation, such transitions are remarkably stable. It is only when conditions of quasi-stationarity are approached -- in the saturation phase of the logistic growth trajectory -- that structural instability and associated short phases of structural discontinuity emerge" (Foster and Wild 1996:252).

In order to guide our impact analysis, we need an analytical or conceptual framework. Such a framework is not a theory. As Leibenstein explains (1976:17), a theory is " a set of relations that are sufficiently specified so that some conceivably falsifiable conclusions can be reached". An analytical framework is a much looser construct: it is a set of relationships that do not lead to specific conclusions. For instance, a simple set of supply and demand functions where both demand and supply are a function of price is an analytical framework. One would need to be much more precise to be able to generate falsifiable conclusions. A framework is helpful in organizing our thinking in a preliminary way.

A useable conceptual framework to look at the impact of innovations on a socio-economy must have a number of important characteristics.

First, a conceptual framework to examine the transformation of the socio-economy in the information age must be able to accommodate a variety of alternative explanations. It would be unwise for instance to presume that the technology-society nexus is a one-way street. So even though we are interested in the impacts of technological innovations on the socio-economy, we must also accommodate the possibility of a feedback mechanism (from the socio-economy onto the ionnovation system) that might at times be the driver in this socio-technical dynamics. It would be unwise to presume that technological innovation is an autononous and exogenous variable, when in fact, as Williams and Edge (1996) and Nico Stehr (in this volume) remind us, some socio-economic forces may indeed shape the innovation process.

Second, whatever might be the fundamental impulse generating the wave of technical change, a useful conceptual framework must be able to identify the channels of transmission through which innovations impact on the socio-economy. These pertain both to the pace of the penetration of the impacts of the technical advance but also to the diversity of complementary channels through which these impacts flow, and the factors that may influence the rhythm of propagation.

Finally, a useable conceptual framework must be able to suggest a categorization of the consequent socio-economic impacts at the micro, meso and macro levels, and to spell out the net effects at each of these levels on the different performance measures in good currency.

High performance may obviously take different meanings. Some are centrally concerned about productivity while others are keenly focused on equality of income; many are interested in work organization, process efficiency, and the organization of production while others are mainly concerned with power shift, agency and social dislocation.

The best way to approach this nexus of problems is to decompose them into families of issues: those pertaining to structural dimensions *and* those pertaining to behavior, process, knowledge diffusion and collective learning.

At the structural level, it must be recognized that the interfaces between the innovation process and the socio-economic process will depend in a significant way on the broader socio-cultural context within which they are embedded.: the more solidaristic the values and the more high-trust the social context, the more these interfaces will prove capable of generating the requisite degree of cooperative and associative governance, the better the fit between the different types of structural capital, and the more effective the mediating agencies underpinning social learning (Hollingsworth 1993).

But there is also a need for some coherence between the structure of the innovation system and of the broader socio-economic system: a misfit between the two can only lead to much friction and poor collective learning. Organizations like the firm are information systems. They are defined by their capacity to transmit and stock information. Innovations in ICT have reduced dramatically both the costs of information transmission and the cost of information stockage. This entails in the first case a reduction in the cost of monitoring, with the result that one can much more easily outsource certain services that are not part of the core competencies of the firm. In the second case, this limits considerably the importance of economies of scale: the optimal organizational form will call for much outsourcing and downsizing. To the extent that mediating agencies (like the firm) do not adjust their organizational form to fit the new circumstances, there may be much waste.

Dudley (1996) has shown that the informational innovations call for the optimal organization form of the firm to become more horizontal, but that different national economic regimes have reacted quite differently to these technological imperatives. In West Germany, the adjustment has been quite rapid, while in East Germany the persistence of the vertical firm has entailed a significant economic retardation.

Consequently, both the socio-cultural milieu and the nature and flexibility of the mediating agencies will be determinant for the impacts of informational innovations. When these two sets of basic parameters (the first ones, the socio-cultural framework, are more or less a given in the short term, but the second ones, the mediating agencies, are clearly a matter of choices in the medium term even though these choices will obviously depend to a certain extent on the socio-cultural framework) are well adjusted to the evolving circumstances the impacts will be important on productivity, and high performance will ensue. When these families of parameters are perverse, there will be little effort to take advantage of the informational innovations, and the general economic performance and competitiveness will decay.

To synthesize this image of the overall structural interdependences defining the interfaces between the innovation system and the socio-economic system, one may use the schema in Table 1.

At the behavioral level, there is much ambivalence when it comes to ascribing causality to historical and contextual circumstances. So many contradictory forces are at work that often one cannot generate any general analytical solution. Yet, the innovation process and the socio-economic process depicted in Table 1 are not deterministically defined by their structures. They are defined also in a loose way by a nexus of broad environmental forces, by local circumstances and random factors, but also by the logic of these processes.

Consequently, it is very difficult to specify what sort of behaviour will prevail de facto in either system, at any given time, or how innovations will impact on socio-economic behavior. Quite clearly, informational innovations increase dramatically the range of possible behavior of consumers and the range of potential business strategies of firms. But the parallel forces that make possible globalization and centralization of decision-making, also, at the same time, allow moves toward localized communities of practice and decision-making. It is difficult to determine which forces will prevail *hic et nunc* (Paquet 1999).

The central role of mediating agencies in the determination of behavior has not been analyzed well. Yet they play a crucial role and represent key policy parameters since they can more easily be transformed than the broad sociocultural and socio-economic parameters. But even the features of these mediating structures cannot provide a way to forecast behavior or to prescribe policy action. For fundamentally, what is the sinew of this whole set of interfaces between innovation system and socio-economic system in

Table 1. Structural dimensions of the interfaces between innovation systems and socio-economic systems

Socio-cultural framework

Innovation System	**Governance Mediating Agencies**	**Socio-economic System**
Cognition	**Organization forms**	**Micro-level**
	Social capital	New value chain
Interactivity	Structural capital (systems, structure, strategy, culture)	Shorter cycles External linkages
Reduction in cost of transmitting and storing information	Intellectual capital (human, customer and structural Techno-organizational forms of firms	Localized and decentralized decision making
		Meso-level
Dispersion	Cognitive arrangements	Networking
	Relational contracts	Work organization
Expanding networks		Disintermediation and new intermediation
Tacit knowledge	**Behavioural patterns**	Development blocks
	Communities of practice	
Self-organization	Trust and confidence	**Macro-level**
	Corporate culture	Income polarization
Interdependence	Values	Higher productivity
		Stabilization
Connexity	**Techno-economic dynamics**	Demographic features
	Network externalities	
	Standards	
	Informational externalities	
	Switching costs	
	Social learning cycle	
	Connectivity	
	Receptivity	

an information age is the process of learning. And learning may be stimulated or blocked by a wide variety of factors.

To analyze the learning interfaces, Max Boisot has used a three-dimensional space – *the information space* – which identifies an organizational system in terms of the degree of *abstraction*, *codification* and *diffusion* of the information flows within it. This three-dimensional space defines three continua: the farther away from the origin vertically, the more the information is codified (i.e., the more its form is stylized and simplified); the farther away from the origin laterally eastward, the more widely the information is diffused and shared; and the farther away from the origin laterally wesward, the more abstract the information is (i.e., the more general the categories in use) (Boisot 1995).

To illustrate the use of the information space, Boisot has identified a few transaction arrangements corresponding to different loci. First, he identifies

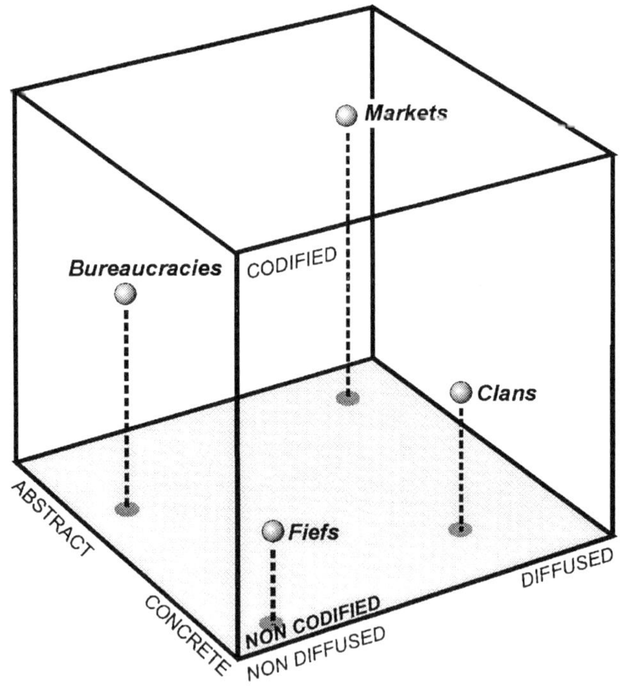

Figure 1. Ideal types of organizational forms

Source : Max Boisot, 1995, pp. 237, 190.

as a fief the type of organization where information is very concrete, and is neither much codified or widely diffused because of the fact that the core transactions are based on the savoir-faire and personal authority of the leader. Second, Boisot characterizes the world of bureaucracy: information is more abstract (monthly reports, etc.) and codified in precise rules, but available only on a need to know basis within the organization and therefore not shared much. Third is the market organization, where price information is very abstract, highly codified and widely diffused. Finally, Boisot labels clans some organizations based on shared values and personal contacts: in the clan, information is concrete, non-codified and widely diffused. These ideal-types are mapped in Figure 1.

In each of these sub-spaces, governing and governance connote rather different realities. In fiefdoms and bureaucracies, governing is based on top-down command and governance is rather hierarchical; in markets and clans, governing is much more a lateral endeavor, and the pattern of governance much more horizontal and transversal.

Within the cube, Boisot has attempted to stylize the operations of the social learning cycle to capture the different phases of the processes of production and diffusion of information in organizational learning. This cycle is presented in two phases with three steps in each phase: phase I emphasizes the cognitive dimensions of the cycle, phase II the diffusion of the new information.

In phase I, learning begins with some scanning of the environment and of the concrete information widely diffused and known in order to detect anomalies and paradoxes. Following this first step (s), one is led in step 2 to stylize the problem (p) posed by the anomalies and paradoxes in a language of problem solution; the third step of phase I purports to generalize the solution found to the more specific issue to a broader families of problems through a process of abstraction (at).

In phase II, the new knowledge is diffused (d) to a larger community of persons or groups in step 4. Then, there is a process of absorption (ar) of this new knowledge by the population and its assimilation so as to become part of the tacit stock of knowledge in step 5. In step 6, the new knowledge is not only absorbed but has an impact (i) on the concrete practices and artefacts of the group or community.

Boisot has stylized this learning cycle in Figure 2, but has also noted that corresponding to each step are possibilities of blockages of the learning cycle: in Phase I, cognitive dissonance in (s) may prevent the anomalies from being noted, epistemic inhibitions of all sorts in (p) may stop the process of translation into a language of problem solution, blockages preventing the generalization of the new knowledge because of the problem definition being encapsulated within the *hic et nunc* (at) may keep the new knowledge from acquiring the most effective degree of generality; in Phase II, the new knowledge may not get the appropriate diffusion because of property rights (d) or because of certain values or very strong dynamic conservatism which may generate a refusal to listen by those most likely to profit from the new knowledge (ar) or because of difficulties in finding ways to incorporate the new knowledge (i).

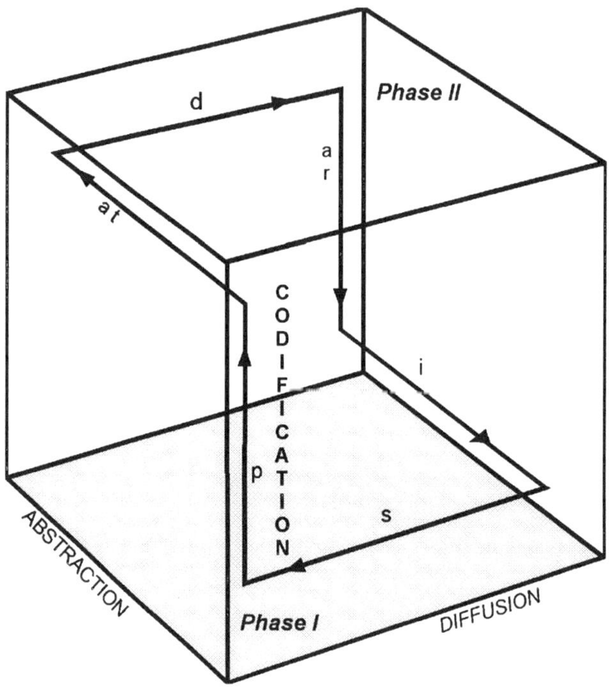

Figure 2. Learning cycle and potential blockages

Source : Max Boisot, 1995, pp. 237, 190.

It is important to note that the social learning cycle does not pertain only to the search of new means to reach well-defined ends. It is a double-looped learning cycle in the sense of Argyris and Schon, i.e., as the learning proceeds, anomalies and paradoxes are generating the redefinition not only of the means but also of the ends (Argyris and Schon 1978).

On the innovation system side, we need not probe any further for the purposes of this paper. We will take informational innovations as a given, obviously influenced by the workings of the broad socio-cultural forces but also of the dual dynamics of the innovation process and of the socio-economic process. On the socio-economy side, there are some micro, meso and macro performance measurements that will serve as benchmarks to determine the nature and importance of the impacts of informational innovations. But there are also broader socio-cultural impacts of some of the technological change on the overall fabric of the social context. Sociologists have shown how some innovations like the bicycle or the car have transformed the mores of modern societies. Finally one also need to pay much attention to the learning cycle in the socio-economy, to the ways in which the mediating agencies embody the learning process, and to the impact of informational innovations on the learning cycle per se.

In all this, the mediating agencies are at the interfaces between the innovation system and the socio-economic process and help us understand the channels through which the informational innovations impact on the socio-economy.

The first family of mediating structures are the *organizational forms* (firms, structural capital, social capital à la Coleman (1988), etc.). These structures, which are all information systems of sorts, are affected directly by improvements in the technology for transmitting and storing information. To the extent that these forms are "creatively flexible", in the sense that they can easily transform themselves to take advantage of new opportunities, the impact of innovations on the socio-economic performance is likely to be significant.

As shown on the right side of Table 1, this will translate into new value chains, shorter product cycles, more localized decision-making, more extensive networking, some disintermediation but also the appearrance on the scene of new brokering services. The new centrality of business services is probably one of the most important organizational changes. Indeed, Antonelli (1998) has analyzed the co-evolution of ICT and knowledge-based business services in Europe, the former enabling the rise of the latter, and the latter acting as a linchpin in the innovation system between codified/semi-codified knowledge in the firm's external environment, and tacit knowledge contained within the firm. Antonelli's work notwitstanding, since we have so much difficulty in measuring services, the rise in intermediating agents can only mean additional difficulties in assessing the socio-economic impact of innovations.

The drift away from bureaucratic regimes of governance toward more market or clan-like forms will mean the emergence of more horizontal management structures and a greater reliance on management by projects. This in turn will inject a degree of nimbleness in the socio-economy and a degree of precarity in the socio-cultural context that may increase significantly the degree of anomie, social dislocation and distrust.

The second family of mediating agencies are the *behavioral patterns* defined by values, degree of trust, corporate culture, communities of practice, etc. This connotes the pattern of values in good currency and the capacity to build on them to make the highest and best use of relational exchange or not. Such capacity to cooperate economizes greatly on transaction costs, serves as a guarantee against opportunism, and may provide, when appropriate, a heightened capacity to take advantage of the new opportunities generated by the informational innovations.

In the same way that technological innovation triggers organizational change, it affects value change. People share certain values and typically make judgments and choose to act in ways that reveal and reinforce these values. But choice behavior is obviously influenced by the options available. When options are expanded or changed by technological advance, behavior can in turn be expected to change, given appropriate lags, and to become "habitualized" into a changed set of values (Mesthene 1970:50).

Either the behavioral pattern will turn out to be rigid and pliable to a very limited degree, and the technological change will translate in simple denial and rejection, or it will turn out to be pliable and very adaptive, and the value change that ensue will not only impact on the socio-economic context but on the whole apparatus of values and beliefs that underpin the socio-cultural framework. In the latter case, the impact of technical advance is made extraordinarily stronger by the existence of a supportive echo box of values and beliefs that encourage the exploration of the new options that have been opened by technological change.

A third set of mediating agencies can be found in the *new market dynamics* brought about by advances in ICT. These novel characteristics change the coordination mechanisms in the marketplace through information externalities, network externalities and important switching costs.

Information externalities, brought about during the adoption phase of a new technology, have important consequences on firm behaviour. As suggested by Shampine (1998), when firms can costlessly monitor a new technology's performance by watching how a competitor fares with the product rather than adopting it themselves, too little adoption may occur.

With respect to network externalities, sub-optimal outcomes may also ensue in technology adoption when technically inferior products become the industry standard simply because the majority of users have adopted them. Tipping of the VCR market to VHS technology rather than Betamax is one such example. Choi (1997) examines this sort of phenomena, postulating that the "second-mover" selecting between two alternative technologies, one which is already being used by a competitor, and another which has not yet been adopted by any firm, will prefer to adopt the product already in use. The risk inherent in adopting the non-tested product is that if it turns out to be inferior, the rest of the market will choose the other technology, leaving the second-mover outside of the network.

New advances in ICT can lead to important switching costs for those migrating to the new product. As each subsequent generation of technology becomes available, expert users of the present technology may be unwilling to undergo the temporary drop in productivity necessitated by a switch to the new product. While in the short term, failure to migrate may yield a temporary competitive advantage through avoidance of costly learning curves, in the long term, this user "stubbornness" can result in no long-run growth (Jovanovic and Nyarko, 1996). Switching costs highlight the importance of a socio-economy's willingness to learn and change, suggesting that economic growth will be dampened in the presence of rigidity towards change.

The fourth family of mediating agencies (very closely linked to the other three but separate from them) are contained in the *social learning cycle.* They are more closely related to cognition and interaction and pertain to the capacity to learn of the socio-economic system. This depends obviously on a variety of features of the socio-cultural framework, on the set of underlying values and beliefs, on the pliability of organizational forms and the amount of tacit knowing they promote, and on the nature and importance of the existing blockages to cognition and to knowledge dissemination within the socio-economy.

The percolation of advances in ICT throughout the socio-economy is affected to no small extent by the *connectivity* (connections within and without) and *receptivity* (openness to learning) of an organization (Antonelli, 1996). Blocks to connectivity and receptivity can stem from organizational strategy, structure, and culture, and, as Nix and Gabel (1996) point out, ideological presuppositions based on organizational values, belief systems, and guiding principles.

This family of mediating agencies is much more complex than the other three. It is nested at the interface of the innovation system and of the socio-economic system, but it depends fundamentally on the social ethos and the goodness of fit among the different sub-systems (technical, social, economic, political, etc.). At the core of social learning is a fundamental capacity to transform that depends much on the connexity and interactivity of the societalcontext. Speed and creative flexibility and ada^ptability in learning will depend on a very strong ethos or sociality to support it.

These four sets of governance mediating agencies (organizational forms, behavioral patterns, tehno-economic dynamics, social learning capabilities) are embodied in a rich "social software" that may act either as a multiplier of the technological impact or as a stopper or attenuator of this impact. Studying the simple correlation between innovation system activities and socio-economic echo effects cannot suffice. It can only lead to naive and deceptively misleading conclusions. Yet, we are condemned to such analyses as long as the body of evidence (statistical or not) is not focused on governance mediating agencies.

Indeed, as John Elster (1989) is fond of reminding us, "explanations" in social science often do not go much beyond assertions, correlations or story-telling. They talk about what may have happened without specifying the "mechanisms", the ways in which things happen. In the social sciences, mechanisms are what matters. So exploratory studies, data collection, and analyses should focus much more on increasing knowledge about ever-more mechanisms. Several mechanisms may lead to the same outcome, so for prediction purposes, there may be no need to distinguish among them. But for purpose of explanation, mechanisms are of central importance. This is the reason why in trying to understand the impact of technological changes on the socio-economic system, an emphasis on mechanisms is pivotal.

There is no way in which one may hope to provide a taxonomy of the major structural, behavioral, market dynamics, and learning mechanisms at work at the interfaces between the innovation system and the socio-economic process. The main reason is that the confluence of forces at work makes each particular set of circumstances rather unique: rigidity of the values system, cultural blockages, learning disabilities, high switching costs, difficult percolation, etc. -- it is the nexus of such factors and many others that are responsible for an accelerated or a stunted impact of some technological advance.

The study of the socio-economic-cultural impacts must however rest on a broad map of the terrain and this is what Table 1 and Figure 2 have attempted to piece together. What has not been presented clearly enough however is the way in which these sets of processes mesh and the exact role of the governance mediating agencies.

One might best state the nature of the challenge in the following way.

Innovations in the information age need to be ingested by the socio-economic system, and this can only be done through some form of learning (absorption, codification, diffusion). This learning occurs at many levels simultaneously (individual households, firms, networks, broad-based systems) and is sustained by different mediating agencies (values system, ideologies, corporate culture, communities of practice, social capital, etc.) which may either provide boosters or blockages to learning and diffusion.

The focus on the processes we have called mediating agencies is of the greatest importance, for any attempt to by-pass them will simply allow the analysis to degenerate into mindless correlations between features of the innovation system and features of the socio-economic system. But this work is extraordinarily complex. Modeling processes is difficult, and data collection pertaining to processes is quite complicated. And even when this is done (as in the work of John Baldwin (1995)) it may not suffice without a new lexicon and new frameworks and measurements being developed.

This is the reason why so many astute analysts, like Leonard Dudley or Ian Miles, have attempted to tackle the problem in an oblique way through simple indicators capturing in some rough way complex adjustment processes. Such ingenuity is to be praised but is condemned to be of use only up to a point.

The Framework for a Statistical Information System (Statistics Canada 1999a) is obviously a crucial effort to define more closely the key linkages between activities and outcomes.

But there is a fundamental need to become more adventurous on this front, for "the slenderest knowledge that may be obtained of the highest things is more desirable than the most certain knowledge of lesser things" (Aquinas). By highest things here, we mean processes. This means that there must be an effort to tackle directly through surveys the process dimensions and to imagine useful measures of these processes.

Indeed, one of the objectives of the research workshop has been to make some progress in this direction.

6. OUR RESEARCH STRATEGY IN THIS BOOK

The research workshop, where many of the contributions to this book were originally scoped out and discussed, was not organized with the objective of resolving all the important challenges we have just reviewed. The workshop was designed to serve as an exploratory vehicle to help Statistics Canada think through what Susan McDaniel has called "the elusive social impact of technology". In the next sections, a variety of avenues and research strategies are explored.

First, Ian Miles underlines the importance of the service sector as a nimble interface which registers in a most fascinating way the impact of technology on the socio-economy. Very much in line with the analysis of Dudley (1996) and Antonelli (1998), Miles suggests that the impact of the new ICT will translate into the emergence of new knowledge-intensive business services which will provide firms with much needed outsourced services. In this way, business services serve not only as a mechanism to make the firm more nimble, but serve also as facilitator in the process of adaptation of the socio-economy to the new technological possibilities.

Second, Keith Newton follows the same line of reasoning, but underlines especially the transformation of the organization and the evolution of the intellectual capital of the organization as the conduits through which ICT impacts on the socio-economy. He also mentions the importance of some of the features of the cognition process as providing an indication of the greater or lesser capacity to transform of organizations.

Third, Graham Vickery pursues the same general strategy but focuses mainly on work organization as the way in which high performance workplaces take advantage of the new technologies. Vickery also probes the ways in which experimentation and effective diffusion of effective work arrangements may be fostered.

Finally, Susan McDaniel, very much like the other three and yet more starkly, emphasizes the central importance of "linkages" (a word very much echoing our call for attention to "mechanisms") between technology and society as the crux of the matter. She also sketches a research strategy that would make use of existing data sets, but would mostly emphasize new frameworks, longitudinal studies, and new measures of process concepts.

The central contribution of this section is to put at the core of the new measurement requirements the need to focus on process and on mechanisms linking technology and society. Each of the authors has approached the problem from a slightly different angle, but they all have emphasized some particular mechanisms (the rise of business services as transformers, intellectual capital and cognition, experimentation and diffusion of new work arrangements, new perspectives both on the conceptual (process) and methodological (longitudinal studies) fronts.

In the following section, the gap between the sort of data that need to be collected and what sort of data are in fact collected is illustrated in a stark way. It is not that the data collected are uninteresting. Rather, it is that we still lack the sort of conceptual framework that will take us beyond the simplistic examination of comparative degrees of penetration of new technologies, and simple inter-regional comparisons of types of uses of new technologies.

The contributions from George Sciadas, Daood Hamdani and Georg Licht correspond to the best available data. They have provided useful measures of technology dissemination at the level of households and in the nexus of firms in the financial sector in Canada, but also the impact of innovation of skills throughout the service sector in Germany. Yet, one must recognize that what we know is limited vis-à-vis what we should know. There has been little attention given to process: we have not progressed yet much beyond the distillation of plausible correlations.

The unsatisfactory state of affairs in the data world is hardly surprising. Without the new templates that Susan McDaniel calls for, little can be expected from data gatherers. What is important is the development of new process-based perspectives if useful data gathering is to ensue.

In the fourth section, impacts of technological changes are analyzed in a more sweeping way through an examination of four specific impact zones: the multinationals, privacy and property, telelearning, and the productivity paradox. In each case, the intent is to provide a genuinely new look at four different "linkages": the multinational conduit, the legal channels, the learning institutions, and the sort of paradoxes leading one to suggest that maybe the cause-effect relationship innovation-society may have to be shifted around.

The key lever in Cliff Wymbs' contribution is e-commerce. It is suggested that it will completely transform the whole socio-economic order. Valerie Steeves is showing that the legal order is probably ill-suited to take advantage of the new innovation-driven system, so she is led to suggest a reframing of the privacy/property foundations of modern democracy. Laura Winer shows how education has been transformed by the new technology. Nico Stehr is more fundamentally radical. He attempts to show that the causal relationships may well not be from the technology to the social, but vice-versa. And John de la Mothe traces not only the socio-historical frameworks that have contextualized our contemporary perspectives on politics, public management and public policy, but calls for a new adaptive framework for governance in an information age.

Each of these five papers is somewhat revolutionary. The conduits they underline (the disturbing forces of e-commerce, the conservative forces of the privacy/property legal order, the stifled and institutionnally paralyzed educational order, and the tectonic change in the social order that may have forced technology into a reactive mode) are triggering unforeseeable changes which make the simplistic tracking down of innovation-society impact look rather silly.

The last part of the volume explores a broader canvas in search of new directions for public policy, research and measurement in such a complex information age. Whether the disturbing forces can stand in the way of the technology-society thrust and derail it in any way is an open question. In this turmoil of forces, adaptation mechanisms are clogged with delays, erratic dynamics, bizarre percolation processes, and important switching costs. Power fields are real, traditions are real, learning is slow, structures are rigid, and behavioral patterns are paralyzed by value systems that are impeding the process of change. This is not as a result of insufficient analyses but rather as a result of immense complexity.

Consequently, the task will not be over when new frameworks will have been designed and new data collected. What will be required is a true transformation of our notions of public policy and public engagement in a world where we are slowly drifting from a focus on government to a focus on governance in our public policy-making.

---

[1] The most valuable contribution of Monica Gattinger is gratefully acknowledged.

[2] The first two sections of this paper borrow freely from the introductory chapter of J. de la Mothe and G. Paquet (eds.), *Challenges Unmet in the New Production of Knowledge* (PRIME 1997), a document that was produced to present and contextualize the Third PRIME Lecture Series.

PART II

ORGANIZING

Chapter 2

Interactivity and Intangibility

Another Pair of I's

IAN MILES
CRIC and PREST, University of Manchester

1. INTRODUCTION

Information is not knowledge. Knowledge is often defined as organised information, just as information is organised data. Data are usually organised by human classification and correlation of observations, but there is increasing capability to achieve this automatically. For example, sophisticated process control equipment can respond to changes in a set of parameters, or neural network based alarms can discriminate between, for instance, Closed Circuit TV (CCTV) images of intruders as opposed to those of windblown leaves. Knowledge attaches meaning to information. It demands a knower, with cognitive structures that can do this.

These structures may themselves be transformed. This may happen unconsciously, or with (partial) conscious awareness as human agents reflect on their knowledge, "model" elements of their own cognitive structures, and sometimes recognise their limits. In processes of interactive learning it is common for participants to develop mutual understanding – awareness of each other's cognitive structures. Agents may seek to learn from or influence each other, thus modifying their understandings and changing the potential for creating new knowledge from information.

Knowledge, then, is a human practice, rather than a thing that resides in artefacts. Thus much talk of "embodied knowledge", "knowledge flows", etc. is rather loose. Knowledge may indeed be shared between people, but this is a process of learning about each other's knowledge, learning to develop common cognitive structures. Sometimes this knowledge sharing may be carried out to facilitate useful exchange of information: for the "source" to be able to translate information so that it is assimilable into the "receiver's" cognitive structures, or for the latter to modify their understandings so that they can make good use of the information supplied.

Rendering information useful requires that it can be produced and presented in ways meaningful to the user, and accessed by and delivered to the user in a timely fashion. The burden does not just lie upon the information providers and packagers, furthermore. The user also needs to have systems for understanding and acting upon information that are able to recognise its salient features. In other words, *interactivity* is required, so that there can be a process of mutual recognition of requirements and opportunities – often this is a learning process. Indeed, we can see that the sender-receiver or source-user model may be a good starting point for basic information theory, but is very limited where it comes to thinking about the construction and application of knowledge. The roles of source and user are liable to be frequently interchanged, and both parties can be thought of as participants in the joint production of a "knowledge space".

This chapter will examine the significance of interactivity in our economies, first through an analysis of the roles of services and especially Knowledge-Intensive Business Services. Since interactivity is a key attribute of new Information Technology (IT), and new IT is very much involved in the transformations of the service economy, the chapter will also consider the significance of interactivity in IT activities. Traditional media of many sorts are being pushed in the direction of greater interactivity; new interactive services for businesses and consumers are being developed; interactive systems (as in e-commerce) are being used to substitute for human-mediated interaction – or to provide new frameworks for such human exchanges (e.g. *amazon.com*'s book reviews). The chapter considers these issues, and the challenges that will be posed to areas of life that have relied upon human interactivity.

Additionally, the chapter will take up a fifth "I" – intangibility. This is a key feature of many services, and this is one of the reasons that this has become one of the buzzwords of recent economics. Intangibility and interactivity often go hand in hand, but they are separable in many respects. Let us begin by examining the two issues in the light of the growth of the services economy.

2. INTANGIBILITY, INTERACTIVITY AND SERVICES

The rise of the services sector to contribute the largest shares of employment and output in industrial societies is well documented. In terms of the classical definition of a "services" economy as an economy in which more than 50% of these parameters comes from the tertiary sectors, most OECD economies have been services economies for some years of even decades. We would do well to remember the point made by Orio Giarini,[1] that the services economy is more about the role of *service* across the economy, not so much about the rise of specific sectors. Practically all sectors have shown growing shares of white-collar employment, of investment in intangibles, of knowledge-intensity, and so on. Equally significantly, it has become a commonplace of management philosophy – not just rhetoric – to stress that what is being provided to clients is a service, that the particular manufactured product which is the object of a transaction is only of value insofar as it can achieve specific functions for the end-user. Understanding changing user requirements, providing back-up and after-sales, and similar service elements are intrinsic to much manufacturing.

But though the rise of *service* is a generic phenomenon, we can still learn a good deal from the challenges associated with the growth of the services sectors. As the services sectors have grown to dominate the economies of industrial societies, so increasing numbers of researchers and policymakers have begun to pay serious attention to them. Services are still often treated as a "residual" category, which can be understood in terms that were developed to describe manufacturing industry; and as lagging and non-innovative sectors who absorb technologies rather than contribute significantly to the development of new knowledge. But it is now much more widely accepted that such preconceptions about services' role in innovation and innovation systems are no longer valid. Indeed, some of them probably never were applicable to a wide range of services.

Some of the key elements of the new wisdom about services are:

- Many services are in the vanguard of the *use* of technology. Some services have long been major users of specialised technologies – transport services use lorries and aircraft, medical service use surgical instruments and pharmaceuticals, broadcasting and telecommunications use particular types of electronic equipment, and so on. But new IT is being adopted so as to increase *the technology-intensity* of services, which are major users of computer and telecommunications equipment and software. About three quarters of all expenditure on IT hardware in the UK and USA stems from services. Certain types of services (financial services in particular, but also professional and business services) are the major users in practically every part of the world.[2] IT is tautologically applicable to information-processing activities, which are at the heart of the activities of many services, as physical processing of materials is at the heart of manufacturing. Arguably, IT represents a technological revolution for services rather as powered machines did for manufacturing in the nineteenth century.

- The services sector is extremely *heterogeneous* in its use of technology. Some services have long been clearly technology-intensive, some are in the vanguard of IT use - but others are lagging. Earlier we have shown that services whose main functions involve transforming information (e.g. broadcasting, professional services, consultancy) are more intensive IT users than those whose main roles have involved effecting transformations in artefacts (e.g. distribution, waste disposal) or human beings (e.g. personal and health services). Not only do sectors vary a great deal in IT use, but within sectors there is also much diversity. This is notable, for instance, in retail. Here large supermarket chains are vanguard IT users (and pioneers in transport and warehousing systems, and applications of environmental technologies, too); while there is minimal use of even basic systems by small retailers.[3]
- As well as being users of new technology, some services are *innovators* - for their own use in new system configurations and applications, and for wider use when they sell on IT services such as software, databases, and telematics to other clients. It has been apparent, since effort began to be made to gather relevant statistics, that services are significant R&D performers in some countries – carrying out more than 25% of business R&D in the UK and US, for example, and more in Canada and Norway.[4] However, conventional R&D measures still often undercount services' technology efforts. Services typically do not organise their R&D with conventional R&D departments and managers, and often do not recognise that what they are doing in project development is R&D. Furthermore, relatively more of their technology development efforts are devoted to non-R&D functions like training and market preparation.[5]
- Other services also play a substantial role in helping *diffuse* technological knowledge resources, via brokerage, systems integration, marketing, training, and consultancy.[6] We shall see later that these are among the most dynamic sectors in terms of growth of employment and output. There is compelling evidence that they are playing significant roles in transforming the whole economy.[7]

Two features common to many services are intangibility and interactivity. Both are closely related to information-processing, and thus to the application of new IT.

2.1 Intangibility

Some services effect physical transformations, or are embodied in physical media. But even if they do so, the services that are rendered are typically not material things in their own right. In the case of information services, clients are not paying for the physical paper or floppy disc, but for the data or information content. In the case of transport services, one pays for the journey, not the train. Likewise, the service may lease, retail or wholesale a material product, but its contribution is the governance of ownership rights. If we think of intangibility as a parameter, the products of "information services" are further along this parameter than are those of "physical services", but both are much further along than is the norm for manufacturing products.

Associated with intangibility is the frequent characterisation of services as hard to store, transport, or trade across international boundaries. These characteristics are often overstated – for instance, services trade has been growing more rapidly than trade in manufactures, in the European Union over recent years. It is common to say that production and consumption of services are coterminous in time and space, and this has been classically true of many services. But much service innovation is about separating these. One of the most evident features of much technological and organisational innovation in services at present involves efforts to overcome this feature – *to render the intangible tangible, or at least visible.* (The current efforts to account for intangible investments are an interesting and not unrelated parallel.)[8]

Various strategies are apparent here. One is to *add material components* to the service. These include, for instance, the customer loyalty cards, membership cards and similar pocket-sized cards now proliferating in a range of consumer and business services. If smart cards take off, they will no doubt be used widely here. Some of the other material components which are introduced have functionality that enhances the service or its use, some are simply marketing aids – consider the variety of tangible artefacts ranging from instruction manuals, through certificates, to mouse mats and T shirts.

It is often harder to demonstrate the utility of a service than that of a physical artefact, which can be touched or even tried out in the showroom. The demonstrability of intangible services, which is especially a problem where intimate relationships are involved, or where service innovations are being introduced, is thus a challenge. Various methods are used to deal with this challenge, some of them involving innovations in their own right. Examples range from guarantees and quality control certification to demonstration packages (e.g. 'demo' software, shareware, trial periods of use).

Intangible innovations are often poorly protected by existing Intellectual Property Regimes. Patents are oriented to protecting physical applications of scientific and technological knowledge, and have only grudgingly and partially been modified to cope with software and related innovations. Many services are poorly protected, and resort to copyright and other methods – including customer lock-in – to capture the value of their innovations. It is sometimes argued that this is a factor in limiting the innovativeness of services, though there is little concrete evidence for this.[9]

The intangibility of many services – especially, of course, those producing information services – means that information-processing activities are central to much of their operation. This is reflected in the high share of white-collar and the low share of blue-collar employment in almost all services. The information-intensity of the service work opens possibilities for application of IT in service production and delivery, and in the range of complementary service activities that often surround the main product:

Application of IT to service processes is widespread, especially for the large-scale, back-office, number-crunching operations of large firms. The high levels of professional knowledge and craft-like work in many personal, social, and professional services has often limited substantial automation in these cases. This reflects both the complexity of the knowledge that is being deployed, and the ability of the workforce to resist changes perceived as damaging to their interests. (Thus instead of automated systems, "decision support systems" have become the norm – it is only where the volume and pace of activity is so high that we hear of automation like the "program trading" of some financial services operating in money markets, etc.)

It is the information components of service products that can be effectively *delivered* via new IT systems. The emergence of telebanking, -shopping, -booking, -brokering, etc. has effectively freed such services from the constraints of coterminality. No longer is it necessary for supplier to visit client or vice versa. When movement is required, e.g. where there are physical goods (or cash) to be delivered, the client or product transport may be undertaken outside of traditional office hours – what in Japan is called 24hourisation – e.g. home shopping deliveries or calls to the cash machine in the evening.

The core service product may be surrounded by a complex of interactions between supplier and client – in marketing, design, ordering, delivery, payment, back-up, and so on. (It may indeed be hard to distinguish the core service from these interactions.) Since these interactions are often mainly informational processes, there is scope for applying IT innovations. (See below.)

(It should be noted that these points do not just apply to service firms. They also apply to the service elements of activities undertaken by manufacturing sectors. However these activities – like innovation in the service sector itself – have been overlooked in mainstream innovation studies.)

The intangibility and information-intensity of services underpins a line of analysis of services innovation that emerged in the 1980s - the "reverse product cycle" approach.[10] This suggests that IT represents a watershed for services, with most services becoming IT-using services. The argument is then that the services sectors will follow vanguard services, which are already making creative use of IT. Such IT-using services have moved from applying IT to their established practices, to undertaking much more dramatic innovation using the new technology. From process innovations aimed at enhancing efficiency, they move through a period of quality improvement to product innovations: to creating new applications and configurations for this advanced technology. The pattern described could alternatively be characterised as an evolution from *back-office* process innovation, through innovation in the *delivery* of services to clients, to *product* innovation.[11]

The "reverse product cycle" approach portrays the absorption of new IT into services as initially a means of increasing service process efficiency. The services begin by being classically supplier-led in their innovations. IT serves as a *catalyst* for services to participate in constructing their own innovation trajectories. As we shall see later, some services play important roles in facilitating this development across the services economy.

2.2 Interactivity

Another characteristic of many services is that they are "client-intensive" – that is, the service is often designed, produced, and delivered in interaction with the client; and of course the client is involved in consumption of the service, but often the supplier is interacting with them at this moment too. Such interactivity is a major feature of many services, and this may mean involving their clients as more than just consumers. In the case of consumer services, the consumer may plan many details of a holiday package, as well as being highly involved in producing the resulting experience. In the case of Knowledge-Intensive Business Services (KIBS), there is often an early phase of negotiation in which the client and supplier jointly establish the characteristics of the service to be rendered, the problem that is to be solved. There may be co-production of knowledge, with clients and suppliers jointly pooling generic and more local knowledge resources.

Of course, services vary in terms of the importance of the various elements of interactivity, and their have been several helpful efforts to classify different services in this respect.[12] However, the characteristics are common enough to conclude that, together with intangibility, client-intensity contributes to the challenges that confront many services.[13] Where it comes to service innovation, these challenges include those of demonstrating the new services in advance of use, establishing property rights for them, and the need to create trust in the process. As in the case of intangibility, there are innovation strategies that directly relate to the challenges of interactivity.

Likewise, innovation research has developed some specific approaches to services innovation which relate to interactivity. One such uses the unappealing neologism "*servuction*" to describe the penumbra of supplier-client relations surrounding process, product and indeed consumption.[14] In this approach, it has been suggested that innovation in servuction is distinct from, and can fruitfully be contrasted to, innovation in production (i.e. the transformation of raw materials and physical artefacts). However, the application of new IT in supplier-client interfaces may mean that this distinction is less significant than it may have been heretofore.

There are several ways in which interactivity impinges upon services innovation. First of all, there is the problem that innovation will often require involving both parties in the interaction (and often their wider networks too), and thus experimentation may be more difficult than in the case of conventional in-house innovation, and there may be more threats to retention of intellectual property. Second, there are a series of issues raised about innovation measurement. Many innovative activities involving interactions between supplier and client (and networks of clients) are overlooked, under-emphasised, or misrepresented by conventional innovation measurement approaches. Very few innovation surveys offer much scope for identifying, or even mentioning, for example, innovations in delivery processes; marketing; transactions; input of information from clients for choice, design or customisation of the service; after-sales support; and so on. Another problem here for conventional approaches to innovation is the role of customisation. This activity is usually explicitly sidelined in terms of R&D and innovation indicators, since the customisation of a physical artefact was generally seen as minor tinkering with design rather than employing new technological knowledge. But much service innovation is intrinsically entangled with customisation of a continually evolving product. Finally, with the high degree of co-design and co-production of service products, it may be difficult to locate the innovation within service supplier or client. It is not unusual, for instance, for service firms to site their staff within client organisations for periods of time. (Similar problems are raised in collaborative R&D more generally.)

2.3 Intangibility, Interactivity, and Innovation Systems

Many of the issues raised above apply to manufacturing as well as to services. The (intangible) service components of manufacturing industries increase as they become more knowledge-intensive; and (interactive) contacts with clients are increasing (and thus issues of innovation in services within manufacturing are increasingly being posed – even if they do not easily fit the traditional product innovation/process innovation dichotomy). Thus, it is not so surprising that many manufacturing firms are reconceptualising themselves as delivering service functionality, embodied in particular goods but supported by a plethora of knowledge inputs about client requirements, etc. Finally, many of the large-scale developments which are summarised by such (contested) labels as "post-Fordism" and among which we see trends such as smaller firm size, decreasing product life-cycles, and greater product customisation are also bringing manufacturing firms closer to services.

There are, indeed, elements of sectoral convergence: services are becoming more like manufacturing in many respects, with standardisation and/or mass customisation, new organisational and control structures, and more technology-intensity. New IT is used in increasingly similar ways across sectors, with similar working practices. But it would be wrong to see all sectors as becoming homogeneous - considerable differences remain related to the nature of products and markets. While the classic distinctions between manufacturing and service activities that are breaking down, we may expect to see great diversity within each of the grand sectors.

The "peculiarities" of services innovation processes, be they ones concerning client relations, IPR, or the nature of the technical knowledge involved are liable to be experienced across more and more of the economy. Standard instruments for measuring R&D and innovation may be missing out a great deal of what goes on in design, delivery, marketing and organisational innovation. Looking at how service firms manage and assess their innovative activities is important for understanding change in the knowledge-based economy.

Services innovation remains understudied and, in all probability, underestimated even in the more sophisticated analyses that are now becoming available. The record may be set more straight over the coming years, as, for example, results of the 1997 European *Union's Community Innovation Survey II,* which included several (but not all!) market services sectors, become available. Many individual service firms are undeniably highly innovative. But there is reason to think that many services are, for whatever reason, disarticulated from key elements of the national innovation system. This implies that they will often lack access to knowledge, human resources, and networks that could help guide their use and creation of new technologies and organisational practices.

Admittedly, there is little systematic evidence on the location of services in innovation systems; and as usual, what evidence there is testifies to the diversity of services and their roles. However, it is evident that the institutional and informal structures that support innovation appear to be poorly developed for many services as compared to other sectors. The systems for production of knowledge resources are typically less oriented to services. In most countries, Industrial Research Associations, university departments and research groups, and scholarly journals are much more often devoted to manufacturing and its sub-branches than to services. And as we have already noted, systems for protection of knowledge are similarly manufacturing-oriented. The intellectual property systems that have been historically evolved to protect technological innovations, have mainly been involved patents (though software is increasingly covered – uneasily – by copyright, and design rights are also of significance to services).

One source of evidence as to services' linkages with innovation systems comes from an analysis of the UK (Technology) Foresight Programme.[15] This Programme aims to promote networking between innovators and entrepreneurs, and to identify opportunities where government involvement in and/or identification of R&D opportunities, can promote innovation, so as to enhance wealth creation and the quality of life. One element was a Delphi study, attempting to gain systematic inputs of views about technological opportunities and industrial capabilities, social constraints and collaboration requirements. A number of Panels were established, including several in services areas, each of which developed its own Delphi, and pool of people to send it to. Difficulties were noted for the service sectors in identifying and utilising expertise, and so extra effort was put into boosting the sample sizes for these sectors. Nevertheless, the report of the Delphi study[16] confirmed that many services have less access to expertise than do manufacturing sectors. They even have less access to people broadly knowledgeable enough to complete the questionnaire. This may mean that such people simply do not exist, but more probably signifies that they are harder for the Panels to find. Either way these services seem to be poorly linked into innovation processes and networks.

More precisely, the analysis suggested that services that were not traditionally technology-intensive – even if they are now liable to be making use of IT – seem to have more problems in accessing technological expertise, and to have relatively short-term technology vision. In contrast, IT-based information services (like Communications) are more like manufacturing in terms of their access to innovation systems. (Physical services, reliant on motor power and other traditional technologies [i.e. Construction and

Transport] display a mixed pattern: Transport appears to be relatively well developed, while Construction appears more disarticulated, corresponding to familiar complaints about the sector).

The Panel data concerns services that were in one way or another linked into the Foresight process. A survey of UK firms more generally[17] allows comparisons to be made between manufacturing and non-manufacturing firms (the latter set includes utilities and construction, and three service sectors: trade, distribution, FIRE and business services.[18] This sample is biased toward more innovative companies, and the results suggest that even innovative service firms are only half as aware of or linked into Foresight as equivalent manufacturers. Roughly twice as many of the manufacturing firms in the sample than the non-manufacturing firms were aware of but not involved in Foresight (20% as opposed to 10%) or actually involved in the Foresight programme (28% as opposed to 14%).

These are only indications of a state of affairs that requires more thorough analysis. Innovation survey data will provide us with some insights here, since these surveys sometimes ask about sources of innovation and the nature of collaborations that are undertaken. But it has to be confessed that we still have a long way to go before our instruments are really adequate for the job of understanding how networks of firms are interacting and jointly producing innovations.

3. INTERACTIVE LEARNING AND KNOWLEDGE INTENSIVE SERVICES

3.1 Introducing KIBS

The discussion of intangibility, interactivity, and innovation around new IT has led us to thinking about innovation systems and networks. Within such networks, an important role is played by intermediaries, and the topics discussed in earlier parts of this chapter have particular relevance precisely because these intermediaries are typically service firms.. Intermediaries help economic actors identify and utilise appropriate forms of knowledge, and the process of interaction is highlighted. The knowledge economy involves a new infrastructure of knowledge-intensive service firms. The discussion below will focus mainly on KIBS (knowledge-intensive business services), since these have developed much more extensively than any consumer equivalent, and since they are very active agents in the innovation process.[19]

What do we mean by KIBS? These are a subset of all business services, excluding those services to business which are mainly based on blue-collar and/or low-skill work. KIBS typically involve highly trained labour, and the knowledge which their workforce deploys may range from the sorts of understanding of administrative and social systems displayed by traditional professional services like accountants, lawyers, etc., to the understanding of scientific and technological systems displayed by new technology based KIBS (what are sometimes known as T-KIBS) such as computer and telematic services, engineering consultancy, and many environmental services. Interestingly, the more professional services are typically intensive users of new IT, for their information processing and knowledge management tasks, and some of them are highly involved in applying specialised formal techniques. There are many "hybrid" services which combine aesthetic and design expertise with more technical knowledge – e.g. architectural and web design services. And in addition to the sorts of KIBS that specialise in knowledge of social affairs – e.g. market research, forecasting, organisational change agents, and policy analysis services – it is common for KIBS to require knowledge (or capability to develop knowledge) of human and social factors affecting their users and end-users.

After long neglect from statisticians and industrial analysts, business services and KIBS in particular are beginning to receive attention. Statistical and policy analyses are emerging, such as the European Commission DG3) report from which the classification of business services in Table 1 below is drawn.[20] Even this classification omits some important KIBS, most notably, perhaps, NACE class 73 (research and development), but also a whole range of telecommunications and financial services. (A problem with the latter is that the sectors involved often serve both business and consumer markets.)[21]

Table 1. E urostat Classification of Business Services

NACE Rev 1 codes	Broad category	More specific services
72.1 – 6	Computer	**- Hardware consultancy** **- Software consultancy** **- Data processing** **- Data base activities**
74.11, 74.12, 74.14	Professional	**- Legal activities** **- Accounting and tax consultancy** ***- Management consulting***
74.13, 74.4	Marketing	**- Market research** **- Advertising**
74.2, 74.3	Technical	***- Architectural activities*** **- Engineering activities** **- Technical testing and analysis**
71.1, 71.21–23, 71.31–33	Leasing and renting	- Renting of transport and construction equipment *- Renting of office machinery incl. Computers*
74.5	Labour recruitment	*- Labour recruitment and provision of personnel*
74.6, 74.7	Operational	- Security activities - Industrial cleaning
74.81–84	Other	*- Secretarial and translation activities* - Packaging activities *- Fairs and Exhibitions*

Note: Branches that are overwhelmingly KIBS are in bold; obvious T-KIBS are shaded; some branches where T-KIBS may also be active are italicised.

The European Commission analysis notes that in this definition the EU's business services employ more than 11.5 million people (8.5% of total employment) and contribute to more than 15.3 % of value added – more than banking, insurance, transport and communication services altogether (12,1%), and almost three-quarters of that provided by manufacturing. The annual growth rate of both these parameters for business services, at around 5.5% over 1980-94, was well above that of the total economy (0.4% annual employment growth, 1.5% for value added). Like many services, the sector features many small firms (more than 2.5 million) operating mainly in national markets, though trade has been growing rapidly (and more rapidly than that of manufacturing, with EU exports annual growth rate at 8.7 and imports at 10.2 for the period 1984-93, as compared to a little over 5% for both parameters for manufacturing).

These data cover business services rather widely, but omit some services that are very interesting and important KIBS. In an OECD study,[22] a rather sharper focus was obtained in a study of "strategic business services" – taken to include computer software and information processing services; research and development and technical services; marketing services; business organisation services; and human resource development services.

Data were analysed for 21 OECD countries. As with the European study, the OECD analysis reports striking levels of growth in these sectors – higher than those for manufacturing or, indeed, for other services. Around 2 per cent of total OECD employment was accounted for by these services by 1994, with total turnover well over US$ 900 billion (various factors ensure that this is a considerable underestimate). Computer services generated the largest share (30%) of turnover; followed by business organisation services (26%); marketing services (23%) and then R&D and technical services (20%). And, again, average firm size was much lower than that for the manufacturing sector or for the economy as a whole in the countries studied.

Knowledge has of course been produced and used in all economies throughout history, and the sudden interest in the "knowledge-based economy" does not mean that this is in itself something new. What is arguably new is the change in the central role of scientific and technical knowledge, of systems for "knowledge management", and of specialised institutions for generating and aiding the generation and distribution of knowledge resources. The rise of KIBS is one element in this.

Earlier we suggested that many services were poorly linked into innovation systems. However, KIBS are often integral parts of innovative networks. This is particularly true for the new-technology oriented T-KIBS, and less so for more traditional firms in sectors like accountancy and law. T-KIBS often play the role of nodes, and sometimes of orchestrators, of innovation systems, while services more commonly are on the periphery of these systems.[23]

4. KIBS, INTERACTION AND INNOVATION

One important role for T-KIBS is that of *fusion*. They may fuse a variety of different types of generic scientific and technological information, which is dispersed in the economy. But most commonly, they fuse such knowledge with knowledge of the more local requirements and problems of their clients. They develop their own repertoire of understanding of the knowledge resources in the society at large, of the sorts of problems typically confronted by clients of particular sorts, and of ways of bringing the two together.

The demand for such services has a number of drivers. The growing importance of innovation in competitiveness is one such. There may also be a change in the range and variety of resources required for innovations, as generic technologies and local applications build upon an evermore complex body of knowledge.

KIBS are often employed as intermediaries by parties in innovation processes who are confronting various gaps in their knowledge resources – and, sometimes, in their innovation management capabilities. Their role may be one of directly *filling* these gaps, or a more indirect *bridging* one. Filling the gaps involves applying the solution to a problem confronting the KIBS' client. Bridging is in large part a matter of providing tools with which the client can move toward overcoming the problem – by diagnosis of the problem, design of solutions, or supply of background tools or information which the client can bring to bear on the problem.

As for the former, gap-filling role, some KIBS are precisely employed to save their users from having to develop particular forms of knowledge. A task may be out-sourced because it is seen as too costly, specialised, rarely needed, etc, .for it to be worth investing in the requisite knowledge. Facilities management and physical services of various kinds exemplify this.

As for the bridging role, Rush & Bessant (1998) distinguish between several bridging functions of KIBS:

- *diagnosis and problem-clarification*, helping users articulate and define the particular needs in innovation - many user firms lack the capability to understand their problems, or to prioritise them within a strategic framework for action, and outside agencies may assist in this process;
- *brokering*, putting different sources and users in contact across a wide range of services and resources;
- *benchmarking*, where the process of identifying and focusing on "good practice" can be established through an intermediary;
- *expert consulting*, providing particular solutions to particular problems;
- *experience-sharing*, transferring what is learned in one context to another;
- *change agency*, where organisational development can be undertaken with help from a neutral outside perspective.

A related distinction between three roles played by KIBS in supporting innovation in client firms, is that between: facilitator, carrier and source of innovation (Miles et al., 1995).[24] A KIBS firm is a *facilitator* of innovations if it supports a client firm in its innovation process, but the innovation at hand does not originate from this KIBS firm, nor is it transferred (from other firms) by this KIBS firm to the client firm. (Examples include: a management consultant helping a client to introduce a new account management system or developing a new services distribution channel; and a technical engineering firm seconding a team of its engineers to work with the technical engineers of the client to co-produce an innovative solution in e.g. offshore platform construction or subsoil building.)

In contrast, a KIBS firm is a *carrier* of innovation if it plays a role in transferring existing innovations from one firm or industry to the client firm or industry, even though the innovation in question does not originate from this particular KIBS firm. (Examples include: an IT firm implementing and customising advanced and innovative software in a client firm; and a management consultant specialising in CAD/CAM applications helping a shipyard to specify the exact user needs and technical specifications of a new CAD/CAM programme, and subsequently to implement it.)

Finally, a KIBS firm is a *source* of innovation if it plays a major role in initiating and developing innovations in client firms, usually in close interaction with the client firm. (Relevant examples here include: an advertising agency developing and implementing a complete new campaign for a client; a provider of call centre solutions advising and actually implementing a new call centre.) Recent research confirms that T-KIBS can be extremely active sources of innovation. They may conduct R&D not only according to client demands, but on the basis of their own assessment of technological potentials. This is the case for various biotechnology services, for example, where it is often easier for a start-up or spin-off firm to offer services (testing and development of products) than it is for them to go into full-scale manufacture. There are also several documented cases of KIBS acting as orchestrators of firms from various sectors to bring about an innovation process - in one case a consultancy firm took the opportunity presented by a conference that I had helped organise to establish a consortium of firms working on IT in the home, in the "Home of the Future" project.[25] In another case, a technical consultancy was responsible for identifying a specific innovative product, and bringing together the combination of firms required to turn it into a viable market offering.[26]

Some KIBS are involved in solving technologically advanced problems; while few conduct basic research, they are often generating solutions that are privately appropriable for the client. For such services, clients would generally be technologically advanced large companies (primarily manufacturing companies in some economies, though services are prominent users in the UK in particular). More KIBS are, however, dealing with less technologically advanced problems, for which their repertoire of clients is much broader.

In the generation of technological innovations, firms rely on interaction between themselves, sharing learning opportunities and experience, and with other, established sources of knowledge and information. This is because many technological innovations result from combinations of tacit and generic knowledge. The economic importance of generic - that is organised, scientific knowledge, often the result of formal research and development – has long been widely recognised. Tacit knowledge, acquired by localised personal experience and individual learning processes, is now increasingly seen as a major source of technological innovation. An innovative economic environment encourages the development and deployment of such tacit localised knowledge, and enables it to be combined with generic resources. KIBS have a vital role here.

Interaction tautologically involves at least two parties, at least a two way process. So far the interactivity of services has been mainly discussed in terms of exchanges concerning the production, delivery and use of specific services. In the use of KIBS, especially T-KIBS, this two- (or many-) way element can extend to the innovations and preparations for innovation both on service supplier and client sides.

One result of this is that feedback from clients can shape innovations in KIBS firms, just as much as KIBS can influence their customers' innovation.[27] Such interactive learning can apply in the case of some other technology-related services (e.g. those involving wholesale and even retail trade of machinery and equipment.). IT support services, management consultancy, and technical engineering, for example, typically work with their clients in highly interactive ways. Client firms and KIBS providers work together, to understand and find solutions for problems. Through the interaction, the client's knowledge base changes, hopefully rendering them more capable of dealing not only with this specific problem, but with a range of problems that may be encountered in the future. Or, if they cannot deal with them using internal resources, they may be better equipped to know when the problem has arisen and where to go to seek a problem-solver. The KIBS provider also gains more experience from the interaction. At least they learning more about the characteristics of a specific industry, and often they learn about new classes of problem and new strategies for solving problems. The KIBS provider is thus enabled to refine and differentiate the services offered and methods used. Additionally, the KIBS firm should have learned of new business opportunities (specific problems and markets), and – last but not least – have upgraded its portfolio of clients.

So, there is a process of interactive learning. Now, this may in some cases go beyond parties learning about each other, and even go beyond the parties going out and developing new knowledge on the basis of what they have learned about each other's requirements. We have already noted that the production of services is often the result of a joint effort of service provider and client , and that this can e seen as a process of *coproduction*. The quality of the resulting service product depends, to a large extent, on the nature of the interaction between service provider and client, and the quality of the communication process that is involved. But this description applies not just to routine services, but also to innovations – innovations in the service provision itself, and innovations in and/or for the client, as client and KIBS jointly work on solving a problem. There is a *coproduction of innovation.*

KIBS have come to play a central role in transferring and distributing knowledge resources in national innovation systems – and given their rapid pace of internationalisation, in international systems too. They are also, we have argued, in many cases involved more actively in the creation of knowledge, or at least in the combination and fusion of different sorts of knowledge.

They accomplish this directly through their provision of services. They also contribute indirectly through facilitating the mobility of highly educated personnel. KIBS have higher shares of highly educated staff than do most other sectors in national economies. While it is unclear as to whether KIBS feature unusually high levels of labour mobility,[28] recent years have seen a considerable flow of knowledge workers from other sectors into KIBS.[29] KIBS direct and indirect roles have much in common with the roles of organisations within the public knowledge infrastructure – higher education, government laboratories, publicly funded sectoral research associations, and the like.

Bilderbeek and den Hertog (forthcoming) see this as a shift from the traditional public knowledge infrastructure to a second, private knowledge infrastructure of KIBS. Certainly, in employment terms, the knowledge services of T-KIBS are rapidly gaining in importance, while public and semi-public R&D institutes generally have become relatively less important. And with the privatisation and marketisation of some state research agencies and higher education research, there has been a blurring of the boundaries between services offered by the public knowledge infrastructure and KIBS services. Nevertheless, the two infrastructures retain distinctive roles within innovation systems. The story varies from country to country, sector and field of knowledge, but roughly speaking, Universities primarily provide technological resources to large R&D-intensive manufacturing firms and social and administrative knowledge resources to the public sector. In contrast, and perhaps paradoxically, the private KIBS firms have a much broader spectrum of clients, including public authorities and some smaller firms. But generally small firms (and many services other than T-KIBS) have a relatively low level of internal competence, and limited financial resources, and thus lack capabilities for making effective use of KIBS. They and often rely on the very limited offerings available from public or semi-public sources for their external technological knowledge, while large firms and other organisations benefit disproportionately from both knowledge infrastructures.

Computer & IT services, technical engineering and R&D services, and perhaps architectural and a few other KIBS conform most closely to this notion of a rapidly growing "second knowledge infrastructure". Such KIBS are typically highly interactive, transferring and carrying knowledge resources to innovating firms, and primarily operating in a private funding context with highly skilled, often formally qualified, staff.[30] In such fields, public support services are apparently losing ground to commercial T-KIBS – and between the two is often a grey area, where it is not always clear how far T-KIBS activities are being performed in a public or private sphere.

Some KIBS are more likely to be facilitating innovation in the clients themselves than others will. Those services that deal most directly with the client's co-ordination activities or production processes, or with the nature of the product itself are most likely to facilitate innovation. But contributions may come from unlikely quarters: for example, legal services may help stimulate product innovation by providing signals as to products that are more or less likely to fall foul of regulatory or litigation problems. There is considerable work to be done exploring such roles of KIBS in the knowledge economy.

5. INFORMATION, TECHNOLOGY AND INTERACTIVITY

New IT has often been mentioned in the discussion above, but it is important to consider some of its specific features here. As other contributions to this volume make abundantly clear, the capabilities of new IT continue to be enhanced at a dramatic pace. (The limits to the processing power of microelectronic chips seem to be forever being pushed back, so that now Moore's Law is anticipated to apply for at least another decade, maybe two or even more – whereas we were already supposed to have reach these limits by now.) Applications of IT continue to proliferate, and it is significant that the term "ubiquitous computing" is now being bandied around, to describe a future where we have moved from the one-computer-for-many-users (mainframe) and one-computer-per-user (PC) eras, to the many-computers-per-user era.[31] But these many computers are already taking on a diversity of forms – personal organisers, interactive TVs, smart telephones, biosensors, and so on.

These trends help contribute to the increasingly rapid production and circulation of information. The hardware and software, the networks and the applications, have significant roles in shaping the ways in which information is created and used, but are not the root cause of current developments. New IT is making it possible to meet increased informational requirements, in specific ways. These requirements emerge from the long-term development of industrial economies – the evolution of complex societies and organisational forms, the globalisation of social and economic affairs, and so on. The technology does not just pop into existence in response to these requirements, however. It is, as much as any other social trend, a product of conscious effort *and* partial knowledge. It is shaped by - and helps shape - these trends. New technological capabilities open the way for new social practices based on these capabilities, and many of these practices were not imagined by the technology's innovators.

So, there is an IT revolution. While there is much loose talk of revolutions, there is a very strong case that new IT can be precisely characterised as a revolutionary technology. Whether there is an information revolution is more debatable. The case could be made that the radical social and organisational change required to fulfil a meaningful definition of a societal revolution is – at least as yet – hard to establish. Perhaps it is more accurate to think of an information explosion. In any case, the scale and speed of information generation and delivery is being dramatically transformed.

But, as we noted at the outset, information is not knowledge. People often talk about "information overload", and indeed in the 1980s there was even a United Nations University research programme entitled "Information Overload – Information Underuse", whose name signifies the notion that too much information becomes noise, drowning out the meaningful and relevant signals. Massive computer power packed onto chips means that it is evermore possible to construct systems that are respond rapidly to user requests and take account of the specific requirements of individual users, providing something more like human interaction. The real issue is not whether or not a system is interactive - it is how far and in what ways it is interactive.

The term "interactivity" is widely applied to new electronics and IT products, but its use has spread more broadly. (While preparing this chapter, I noticed a range of all-weather clothing in an outdoor clothing shop being described as "interactive".) Now, the specific word may become unfashionable, simply through overuse and appropriation for trivial purposes. But the underlying idea will almost certainly become even more widely diffused, in large part through the incorporation of new IT into all sorts of products. The era of clothes incorporating sensors and communication devices is probably not far off. But IT is not the only factor here – increasing sophistication in our understanding and control of materials means that so-called "smart" materials and fabrics are becoming available. These can, effectively, modify their own properties in response to environmental circumstances. They interact with the physical world of heat, dampness, and the like.

IT systems can interact "smartly" with the physical world too, if equipped with appropriate sensors. But the term "interactivity" more commonly implies interaction with the user, responsiveness to signals from the user that can be interpreted as commands or enquiries. Some IT products' responses take the form of physical activities – actuators that control wheelchairs or artificial limbs, that open doors or pull curtains on command, for example. More often, at present, the output of IT is simply information generated by the product itself, or by the system which it networks into – or communication with other human beings which it is mediating.

Let us consider the cases of human-computer interaction first. At present, much, usually the great majority, of the information processed by the IT system was put there by human beings in the first instance. This is evidently becoming less true as more and more IT systems are relaying information from satellites, webcams, CCTV systems, traffic management systems, and so on. (One of the most visited websites a few years ago was a coffee machine in a US University, whose daily doings were the source of mysterious interest to people around the world. Unless, that is, it was other coffee machines that were logging on…) Nevertheless, most of the time, we are still dealing with text, and secondarily graphical output, that people have themselves created – usually with the aid of PCs, though it is possible to find libraries full of historical literature on the Web these days. Often the text and graphics represent the programmers' informed guesses as to the end-states, which the system will need to report on. (The user's inputs have had such-and-such a result, for instance, or the result of the query is that the value of parameter X is *n*.)

More interesting degrees of interactivity are attained where the IT system puts together information in effectively new ways in response to a user request or stream of commands. Consider two examples:

- *Searching for information resources*. The online database, or the web search engine, comes up with a list or references or websites, compiled according to the search criteria input by the user. Many of our enquiries are probably banal and common ones, but still the response is generated just as originally as if this were a unique request for information.[32] The system has the capacity to assemble available information into an original output, tailored to the user inputs.
- *Playing a video game*. The PC or games console is programmed by the games software to create a virtual world in which various objects, characters and environments are possessed of specific properties. (In

some games there are random and even fractal elements built into these.) The player's choices, moving the protagonist(s) through a series of scenarios or encounters, provide a variety of inputs which lead the system to generate responses – the virtual world responds to the player according to its logic; and the results depend upon the skill and ingenuity of player and programmer alike.

These examples point to what are often very impressive applications of computing power. The first very much corresponds to the notions that people had in the early 1960s about what might be achieved in the future world of high-power, widely-diffused computing, that their technological forecasts lead them to envisage. IT was identified with *information utilities*, and the model was extrapolated from the experience of pioneering users of databases, linked into mainframe machines with their very basic peripherals. (The notion that much of the data might be processed and even stored locally on a PC was a heretical one at least until the mid 1970s.) The second example is a timely reminder that technological potentials are frequently captured by unexpected innovators (e.g. teenagers in the early '80s) who create their own applications. While corporate requirements for financial data (spreadsheets) and word processing spurred the development of the business PC market, the home market was very largely driven by early games applications. The rhetoric about educational uses and other domestic applications (e.g. digital photography) may have a little more substance now, but video games continue to motivate much of the consumer market. (And alongside home computers, there is a huge market – and much competitive rivalry – for consoles, whose software base is almost exclusively games.)

Though impressive achievements, at least when we consider the better databases/search engines and games, these applications are still primitive ones in many respects. I am not so much thinking of the limitations of automated abstracting and virtual reality here, as of the limited learning capacity built into these systems. There is learning on the part of the user, so that search terms or gameplay can become more refined. But the "learning" on the part of the computer system is very limited – signalling where a particular database entry or website has already been inspected, adapting a game situation to the fact that the player has already made certain moves here,[33] and so on.

In this respect, interactive systems still fall rather short of human interaction, where there is scope for interactive learning on both sides of a dialogue. Artificial intelligence routines, of classical heuristic form or perhaps using neural net or other approaches, will gradually come into play. These will render the human-machine interaction more profoundly similar to human-human interaction. We have a way to go here, but already the superficial similarities point to important features of IT systems.

Interacting with new IT is much more like participating in a conventional conversation, as several commentators have noticed. Each message is generated in response to user inputs, rather than following a single sequence predetermined by the source or information supplier. Interactivity places the *flow of information* to some extent under the control of the user. It makes the interaction a two-way affair.[34]

This is quite different from traditional media, where a one-way flow of information has been the typical state of affairs. With mass broadcast and recorded media, and to a lesser extent with print, the sender-receiver model seems to be quite applicable. The user can choose to access the source of information, or not, but has little control over the flow of information. One advantage of print over broadcast media was that, until at least recording equipment was widely available, with broadcasts you simply had to take the input when, and in the order that, the sender chose to transmit it. With print, the reader is free to skip around the text, even to annotate it, to return to passages a second time, and so on. Many people still regard a book as an ideally interactive medium for this reason, and it will take more advances in portable PCs to convince them that electronic media really present a significant challenge here.

The book is still the book, however, with its information presented in a linear form (even if it has a good index!). The reader is the one who has done the learning, navigating around and building a conceptual model of the text. Interactive IT systems may not be able to build models of the user (though wait and see), but they can organise their information in innumerable ways. Often they can draw on several orders of magnitude more text than the average book, too – CD-ROMs already contained whole encyclopaediae, and now DVDs have come along with ten times the capacity – and this is before we go online.

Powerful information processing in the IT system, and capacities for responding to more user input, is underpinning "publications" and “programmes” offering forms of interactivity more like those provided by interpersonal contact. Highly interactive products are often said to be “intimate” ones, with users expected to shape what happens next by being engaged with and responding to a flow of events. There may be “default” modes, in which a linear flow of information is presented if the user takes no action, but the design of the products typically anticipates interactivity.

Another feature that sets new IT apart from earlier technology, and makes its use more like human interaction, is the way in which ordinary language is often used for communication – classically commands are issued, and reports provided, in the form of text. Speech recognition and synthesis gives rise to something even closer to conversation.

New media may also involve new forms of person-to-person (or people-to-people) interactions as well as human-machine interaction. Electronic text and voice messaging systems allow for asynchronous communication and rapid delivery of text. This frees users from some of the limitations of traditional media. Services that can be automated are increasingly handled by human-machine interaction through automated teller machines. Services that require human knowledge may be delivered, at least in part, through communications media.

There is also scope for one-to-many and many-to-many communication in addition to the traditional one-to-one format. Some electronic media allow for contact to take place, anonymously or otherwise, in 'cyberspace'. There may be a virtual environment in which the participants meet, like an office or conference site, or a game environment as in multi-user role-playing games. People may make contact deliberately, or make at least initial contact by virtue of the fact that they happen to be accessing the same service. As on CB radio, on new media such as bulletin boards and chatlines, as well as in the more elaborate game and conference environments, individuals may adopt pseudonyms and even role-play new identities.

Awash in a sea of information, users are requiring assistance – and their needs are at least at present growing faster than the capabilities of agents and similar technologies to cope with them. The forecast that information society would mean *disintermediation* is overstated: while some intermediaries disappear, others are rising. Knowledge brokers of various kinds are among these. Services are emerging to help users locate the information they require; to validate and provide a measure of quality control in the intellectually anarchic Web. Portals have become a holy grail of the browser and Internet Service Provider (ISP) firms. The notion is that users will be loyal to one service or site if it saves them time and effort in accessing websites congruent with their evolving interests.

There is evidently a long way to go before much of the interactivity provided by KIBS can be captured into interactive IT. At present, the main applications of multimedia technology are to enhance the human-to-human interaction, as in the use of IT-enhanced presentations, helplines, etc. The new media also play powerful roles in marketing, and in disseminating codified information – visit the websites of leading accountants, consultants, market research and software firms, for example, to find research papers and position pieces alongside the more straightforward marketing of the firm, presentation of annual reports, etc.

Practically all KIBS employ IT in their processes, and many in their marketing, but relatively few extend this far into their delivery systems. The need for interaction with clients about their specific problems and circumstances is not easily subject to automation. Where a service is fairly standardised, or involves limited customisation (e.g. searching a database according to simple criteria), the client can interact with the service provider's (codified) knowledge base through a Human-Computer Interface. Such applications are common in IT industries, where a software user can often call a dial-up bulletin board or related service, and consult FAQs (frequently asked questions). Often there will be human-to-human communication via email, or telephone, with dialogue between users and software engineers (the latter typically keeping electronic records of the problems and other points raised by users).

Hopes of speedily extending the scope of such systems, and the range of services that they apply to, may be overoptimistic. They run up against the tacit knowledge that is vitally important to many KIBS; tacit knowledge which is an intangible asset, and which often requires high interaction during product design and delivery, and often in the course of use. Whether this is fine-tuning the service at the moment of delivery, or a more major reconfiguration, current IT-based delivery systems mainly serve to augment person-to-person interaction. Trust and confidence are required (which is why many professional services use highly traditional modes of delivery, such as reports and "live" briefings, enhanced only slightly by IT in the form of Desktop Publishing, impressive graphics, and new presentation technology supporting face-to-face briefings). Physical contact and telephony is being supplemented (and perhaps at the margins replaced) by fax, voice and E-mail messaging; videoconferencing and related methods will be an area to watch in coming years.

More advanced applications are most commonly pursued *within* large organisations, where sophisticated "Knowledge Management" systems based on groupware are being deployed. (KIBS may play a supporting role in developing such applications, and some KIBS – mainly large international consultancies – are using them internally.) It will be worth watching these developments, to see how far they extend beyond the boundaries of the firm (for example in collaborative ventures). And, finally, another area to watch is the use of telematics in education and training. Here experience is being gained in generating flexible packages for distance learning. The knowledge created here may be applicable to many more services in the future.

6. CONCLUSIONS

This chapter has examined the rise of knowledge intermediaries of all kinds, especially in the guise of KIBS, and the challenges to conventional wisdom created by the intangibility and interactivity of their operations. We have noted the wide use of new IT in these activities, and the way in which they are often applied in the field of IT innovation in the form of computer, software and telematics services. Finally, we discussed interactivity in the context of the new technology, and the limits to substitution of human intermediation by IT.

KIBS have a highly significant role in the innovation processes around interactive technology. The new technologies facilitate the operation of KIBS, and are used intensively by them. They generate demand for KIBS. And in helping to meet this demand, KIBS in turn are fostering the more rapid development and diffusion of the new technologies.[35] Again, this involves a process of interactive learning. As well as developing new technological knowledge, supply-side innovators are learning about what products and product features are liable to gain market success, to what extent and in which markets, and why. Innovative users of the new technologies are also learning - of the very existence of new knowledge and the capabilities and limitations of products based on it, how different product offerings compare, how to acquire skills for implementation and further development of the innovations.

The process is decidedly dynamic. It is not just an accumulation of knowledge about a static constellation of alternative designs and user interests, and not just an interaction between suppliers and users. As we have stressed, the complexity and diversity of the knowledge requirements for contemporary innovation have prompted the emergence of a range of new intermediaries, offering intangible, interactive services. It will be very interesting to observe how these intermediaries themselves innovate in the use of new IT to facilitate their activities.

Notes

[1] Giarini and Stahel (1993); see also W R Stahel (1998) and other articles in the same issue of The IPTS Report no 27, September 1998

[2] Miles with others (1990); OECD ICCP (1993)

[3] The continuing increases in familiarity, functionality and power of IT, together with the decreases in its price, mean that new generations of IT are liable to be adopted by many of these "laggards". For example, there has been high uptake of mobile communications by many SMEs.

[4] Young (1996)

[5] Cf data on UK services presented in Miles and Tomlinson (1999)

[6] See Miles et al (1995), EC, Luxembourg, (DG13 SPRINT-EIMS), Bessant & Rush (1998).

[7] See, for example, papers by Antonelli, Tsounis and Tomlinson in Miles and Boden (forthcoming), which demonstrate this using input-output data for various countries and periods.

[8] On accounting for intangibles, See Jacquemain (forthcoming) and various papers available on the OECD website, e.g. http://www.oecd.org/dsti/sti/industry/indcomp/prod/intang.htm, which provides a guide to the eralier papers there.

[9] See Miles, Andersen et al (forthcoming).

---

[10] See the studies by Richard Barras (1986; 1990).

[11] For a detailed empirical critique of the reverse product cycle approach, see Uchupalanan (1998). This study provides a detailed analysis of the interplay between competition, strategy, innovation and regulation for a series of banking IT innovations.

[12] Cf Silvestrou et al (1992), who define two extreme categories of service, differing along six dimensions. The extremes are Professional services: organisations with relatively few transactions, highly customised, process-oriented, with relatively long contact time, with most value added in the front office where considerable judgement is applied in meeting customer needs; and Mass services: organisations where there are many customer transactions, involving limited contact time and little customisation, and offerings that are predominantly product-oriented with most value being added in the back office and little judgement applied by the front office staff. A third category, service shops, fall between the two above on the six dimensions. Another approach, from de Jong (1994) distinguishes four groups of services: (1) Infrastructure services, such as telecommunication and transport services, which make use of fixed network facilities for standardised problems. This lacks a parallel in the Silvestrou approach. (2) Value added services, highly specialised business services used by different types of, like accountancy and wholesale trade. (3) Pre-specified services, which make use of a standard approach or method in order to satisfy a general demand, e.g., repair, maintenance and cleaning services. (4) Ad hoc services: for example management, consultancy and engineering services that are called in for specific, one-off problems. De Jong examines the potential for these different types of service to pursue 'economies of scope' and 'economies of scale', which are related to the possibilities for standardisation or variety in the services offered.

[13] And, again, client-intensity is not restricted to services. Manufacturing firms too, vary on this dimension, with mass production being quite different from the creation of a one-off highly complex device for a single client. We merely suggest that it is a marked feature of services, and one that is liable to shape their innovation patterns, and roles in innovation systems.

[14] Belleflamme et al (1986), Eiglier & Langeard (1987).

[15] The following is drawn from Miles, (forthcoming, 1999).

[16] Loveridge et al (1996).

[17] CBI/Natwest (1996).

[18] The sample consisted of 487 manufacturers and 200 non-manufacturers; it was drawn from CBI membership and the mailing list for NatWest's newsletter Innovation Business.

[19] The consumer equivalents to KIBS seem to fall into a few main categories. First are certain charitable bodies, providing advice, counselling, and other professional services. Second are services that are really oriented to microbusinesses (e.g. accountants and IT support), or to people in their roles as investors rather than as pure consumers (finanical advice services). Third are various voluntary associations, for hobbyists (these may involve technological knowledge, as in ham radio and computer clubs) and fans.

[20] Communication From The Commission To The Council (1998).

[21] NACE classes 70 (real estate services), and subclass 74.15 (holding companies) are also not included in this definition.

[22] Vickery and Murphy (1998)

[23] But some other service firms do play "orchestrator" role. Consider the large supermarket companies and chain stores, who can have a considerable influence over their suppliers products and even their processes. (e.g. some large UK food retailers have their own

standards for their agricultural suppliers, and most have been influential in shifting their suppliers into using Electronic Data Interchange systems.)

24 I draw here on examples developed by my colleagues from Dialogic in the Netherlands, Rob Bilderbeek and Pim den Hertog, as provided in their contribution to Workpackages 4 and 5 of the SI4S project. This material is available on CDROM from the STEP group in Oslo, Norway (http:\\step.no).

25 The conference formed an input into Miles (i988), which makes interesting reading today, when much of the technology discussed is on the market.

26 Howells (1999) is a rare study of the emergence of external R&D services. It is useful to read this paper alongside the only specific survey of the sector and its clients that I am aware of, Rose (1997).

27 The literature on user-producer relationships – especially the work of Lundvall (1996). Lundvall and Johnson, (1994) - is thus highly relevant.

28 It has been suggested that this is the case for Sweden, at least (Marklund 's contribution to the SI4S project –see endnote xxiv) but this remains arguable. The small size of most KIBS companies, coupled with what we know about the limited life-expectancy of many small firms and start-ups, would suggest that quite high levels of mobility across employer might be expected.

29 Tomlinson and Miles, 1999.

30 R&D services show a somewhat different pattern; their growth rate is generally slower (though it is particularly high in the UK case). The UK may or may not be the shape of things to come, but at present in most European countries R&D services operate predominantly in the public sphere)

31 See in particular the work of Mark Weiser, e.g. Weiser (1991) – for more bibliography consult the website http://www.ubiq.com/hypertext/weiser/UbiHome.html

32 We should qualify this a little. Some systems do have the capacity to retain search results for future re-use, and some use human selection of particularly relevant sources for specific common topics. Both of these may well be trends that will continue, though for different reasons; but they will be enhanced by "smarter" machine searching systems, as well.

33 One recent computer game features "characters" who act as if they can communicate among themselves about their experience of the player – so that, for example, a player will encounter different reactions as he or she proceeds with the game according to whether they are approaching situations belligerently or in a spirit of negotiation. Video games start to approach game theory, indeed.

34 Which is not to say that there is any sort of equality in it!

35 We concentrate on the mutually reinforcing effects here. Actually, the effects are highly complex, being selective in their own right. KIBS shape certain patterns of technology development, and vice versa. The full measure of this has only dimly been grasped, despite much work on "social shaping of" and "social impacts of" technology.

Chapter 3

Learning Firms

Concepts, Measurement and Policy Issues

KEITH NEWTON
CRUISE, Carleton University

1. INTRODUCTION

The starting point of this chapter is that in the emerging global knowledge-based economy (KBE) powerful forces call for fundamental adjustments by individuals, business firms, institutions of many kinds and, indeed, society as a whole. The literature on the characteristics and underlying trends in the KBE is by now voluminous (Gera, Sing and Newton, 1998). Suffice it to say that, for present purposes, it is the *knowledge-based* character of the KBE, and the adjustment challenges it poses for firms, that are of central concern.

The increased uncertainty, turbulence and discontinuities that accompany the rapidity and ubiquity of technological advances are felt acutely at the level of the firm. Globalization, trade liberalization, increased foreign direct investment and powerful information and communication technologies have combined synergistically to not only intensify competition but also to fundamentally alter its character. To survive and prosper, firms must now commit to continuous innovation --indeed, the expression "innovation competition" has now come into vogue.

Most importantly, the continuous innovation that is required for survival also brings with it another critical requirement --that is, for continuous learning, for the continuous development and application of knowledge. Thus it has become commonplace to aver that knowledge is now the most important factor of production and source of competitive edge. Cognitive abilities --skills, competencies, know-how, knack, flair, creativity, ideas-- are the new stock-in-trade. The monumental challenge for firms and their managers, therefore, is how to identify, motivate and effectively marshal these valuable intangible resources, and profitably apply them. This chapter looks at some approaches to such knowledge production in "learning firms". It is an outgrowth of earlier work (Newton, 1996, 1998) that examined the importance of the human factor in management strategies to improve productivity and competitiveness in the KBE. That work pointed to a nexus of human resource-based management strategies (variants of total quality management, organizational innovation, and the 'new' human resource management) that could work in concert to build high-performance work systems. Two interesting new approaches identified in that work are the focus of this chapter: organizational learning and intellectual capital, respectively.

While the observations about a knowledge-based economy may seem intuitively plausible, it would appear useful to set out some of the empirical evidence to show that business firms and their managers do, in fact, face a more skill- and knowledge-intensive economy and labour market. This is the subject of the next section where a range of recent indicators of growing knowledge intensity is invoked. The third section of the paper focusses on theoretical and conceptual considerations relating to knowledge and learning, and then to organizational learning in particular, and to the notion of intellectual capital. This section addresses definitional and theoretical issues relating to the changing nature of knowledge production, the distinction between tacit and codified knowledge, and the importance of cognitive diversity, "whole brainedness", teams, networks, and interaction, etc. Section three goes on to examine some empirical evidence on learning firms, including very recent work by the author. The weight of evidence leads one to conclude that, despite the plenitude and stridency of the rhetoric, one may concur with the OECD (1996) that here, as with "hard" technological innovation, and as with complementary or "soft" organizational innovation, there is a learning lag.

The fourth section turns to issues of measurement and methodology. It addresses these questions from two standpoints -- from the standpoint of the individual firm but also at higher levels of aggregation that might include the region, industry, sector, or national economy. Some of the policy implications at these various levels are considered in section five, with an emphasis on the role of government and/or specialized, non-governmental intermediaries. This concluding section also attempts to distill major findings and set out priority items for an agenda for future policy research. Some observers of course, are frankly skeptical of the KBE concept.

Box 1: WHY THE NEED FOR LEARNING FIRMS?
WHAT'S SO NEW ABOUT THE KBE?

"What's new", they ask, "the economy has always been knowledge-based". True, it took skill and know-how for our ancestors to successfully hunt food and furnish a cave. But one can argue that the shifts currently in train are revolutionary rather than evolutionary --a new technoeconomic paradigm. To illustrate the fundamental qualitative shift, consider the concept of the production function as set out in basic economics just a generation ago. One learned that "factors of production" including land, labour, capital and (sometimes) materials, were combined by the organizer/manager/ entrepreneur to produce output. Further refinements might allow capital to be broken into physical and financial and (later) human, social and organizational; labour to be split into skilled and unskilled; and technical change could enhance the productivity of all factors collectively.

Now consider the factors in turn. First, the concept of land has changed. ICTs have rendered distance and location less relevant. More and more offices are virtual. Labour, too, has changed. Well within memory the "typical" worker was a male, white, unionized factory worker who embarked on a career of some 35 or 40 years. Now the fastest-growing segments of the workforce are women and recent immigrants and the very notion of a "job" is being questioned. Capital is increasingly intangible. (Think of the relative proportions of tangible and intangible assets in, say, GM and Microsoft). Next, raw materials carry diminishing weight. Think of the amount of sand that goes into a microchip versus the amount of brainpower. Observers now talk about the "dematerialization" of production. Finally, the management function in the production process is also changing profoundly. With the aim of creating a participative learning culture, rigid hierarchical structures are giving way to participative decision-making and interactive teams. The management of brains not brawn is the order of the day. *That's* why we need learning firms.

2. EVIDENCE OF INCREASING SKILL AND KNOWLEDGE INTENSITY

This section collects various pieces of empirical evidence that, together, strongly support the contention that the economy is increasingly knowledge-intensity since this is precisely the environment that is driving firms to organizations that systematically develop and manage their knowledge assets.

First, data from Statistics Canada's Labour Force Survey show that in the recent period of 1990-97 employment growth has been led by highly skilled and educated workers. Jobs for workers with a post-secondary education increased by 34.2% during that period, while those for workers with less than high-school graduation fell by 13.6%. In similar vein Boothby and Gingras (1998) divide the workforce into 2 groups: those with and those without post-secondary certification, crudely referring to them as high- and low-skilled workers, respectively. They find that, for the age group 20-54 years of age, the ratio of the unskilled to skilled fell from 3 in 1976 to less than unity in 1997.

Next, in one of the richest sources of information about skill- and knowledge-intensity in the Canadian economy, Gera and Massé (1996) offer several pieces of related evidence. First they follow the OECD's (1996) breakdown of occupations into white-color high-skilled, white-collar low-skilled, blue-collar high-skilled and blue-collar low-skilled, respectively. They show for Canada how the white-collar high-skilled group has grown significantly in the period 1971-1995

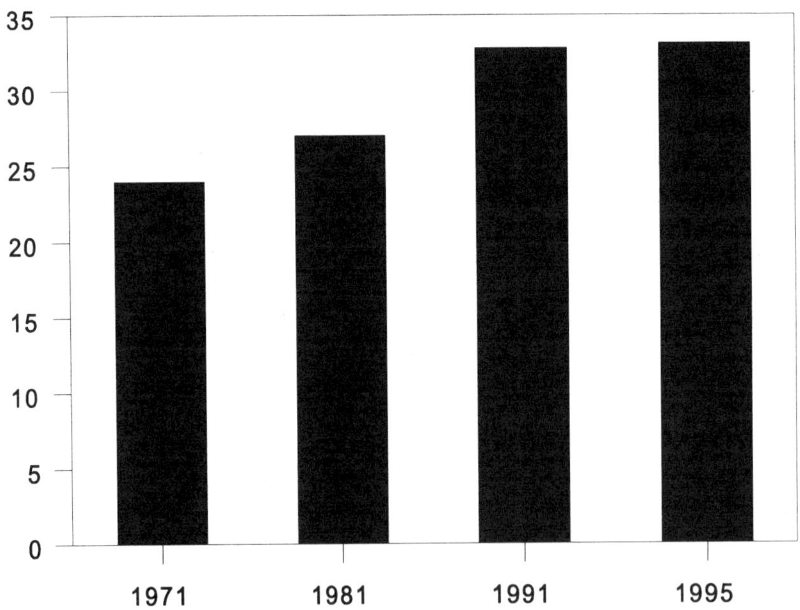

Figure 1. Share of White Collar High-Skilled Workers in Total Employment.

Source Gera and Massé (1996)

Second, building on the work of Lee & Has (1996) they invoke three measures of R&D and three of human capital to rank and partition Canadian industries into high-, medium- and low-knowledge intensity. The three R&D measures are: R&D expenditures; the proportion of R&D personnel in the workforce; and the proportion of university-level R&D personnel. For human capital content the measures are: the proportion of workers with post-secondary education; the proportion of "knowledge workers"; and the proportion of scientists and engineers. To be "high-knowledge" an industry must have two of its R & D indicators in the top one third *and* two of its human capital indicators in the top one third. To be "low" the reverse rule applies. The rest are classified as "medium".

Using these definitions they show that the "high" group grew from 10.7% of total employment in 1971 to 15.4% in 1991. Interestingly, in addition, Table 1 shows that, with an employment share of only 15.4%, the high-knowledge industries nevertheless contributed 41.2% of the *growth* in employment in the period 1986-1991. Low-knowledge industries, with nearly 45% of total employment, contributed only one fifth of the job growth in the same period.

Table 1. Employment Growth by Knowledge Intensity, 1986-1991
Total Business Sector

Knowledge Intensity	**Employment Share (%)**	**Share of Gain (%)**
High	15.4	41.2
Medium	39.7	38.8
Low	44.9	21.5
Total	100	100

Source: Gera & Massé (1996)

Closely related to knowledge intensity at the industry level is, of course, the extent to which this is reflected in the occupational structure of the workforce. When, as in Table 2, information workers are sub-classified as "data workers" (clerical, sales and service occupations) and "knowledge workers" (managerial, professional, scientific, teaching, medicine, social sciences, and artistic occupations) the increasing knowledge intensity of jobs is clearly apparent. Several pieces of evidence point to the observation --of considerable importance in the context of the KBE-- that technological change is skill-biased. McMullen (1996) using data from the longitudinal Working With Technology Survey (WWTS) shows that in response to the introduction of computer-based technologies in the second half of the 80s and the early 90s most jobs created were high-skilled and most of those lost were low-skilled. In addition, all occupational groups experienced an increase in skill requirements. Next Baldwin and Rafiquzzaman (1998) find strong evidence that advanced technologies are complements with more skilled workers in manufacturing.

Table 2. Employment Share of Information Workers by Sector, Canada, 1971-1995

Proportion of Information Workers				**Ratio: Data/Knowledge Workers**[1]		
	1971	1991	1995C[2]	1971	1991	1995C[2]
Goods	22.1	34.3	37.8	2.51	1.01	0.94
Services	59.2	63.8	64.1	6.8	2.93	2.67
All	45.1	57.3	58.5	5.35	2.28	2.27

[1] Data workers include clerical , sales, & service occupations. Knowledge workers include managerial, professional, scientific, teaching, medicine & health, social science, & artistic & literary occupations.
[2] Curvilinear coefficients, COPS 1994.
Source: Economic Council of Canada, *Employment in the Service Economy,* 1991 & calculations by Ekos Research Associates Inc., based on data from COPS.

Finally, recent empirical evidence for Canada (Gera, Gu, and Lin, 1998) uses two skill classification schemes: one developed by the new National Occupational Classification (NOC) and the other following the criteria used in the classic work by Wolff and Baumol (1989). The novelty of this work is that, in addition to documenting trends in skill intensity in Canadian industry, it also investigated the issue of skill-biased technological change, using an index that combines four industry-level measures of technology: R&D capital; the stock of patents used by industry; the vintage of the capital stock; and total factor productivity. Three major findings emerge. First, the relative demand for more-skilled workers rose over the period 1981-1994, and this rise in skill intensity was pervasive across all (29) industries. Second, the shift in demand for more-skilled workers was entirely driven by "within-industry" skill utilization rather than "between-industry" employment shifts -- a result that is consistent with the hypothesis of skill-biased technological change. Third, the various technology indicators are strongly correlated with skill intensity. The authors conclude from the last two findings in particular that skill-biased technological change has been a key factor in within-industry skill upgrading across Canadian industries.

Taken together, these indicators attest to the growing importance of skills and knowledge in the economy. It is worth noting, however, that they have been subject to criticism. Some of the cruder proxies for skill such as the distinction between production and non-production workers (not reported here) clearly mask a multitude of classificatory sins. The same may be said of variables such as level of educational attainment, and experience. Increasingly, attempts are being made to invoke more direct and more specific measures of skill. Thus Leckie (1996) examines occupational trait data from the Canadian Classification and Dictionary of Occupations (CCDO) for the period 1971-91. The traits include specific vocational preparation (roughly, the amount of job-specific training needed for the job); general educational development (the level of reasoning, mathematical and language skills needed); cognitive complexity; and task diversity; all scaled. The important finding is that for all these traits, the employment shares of jobs requiring the highest skill levels rose monotonically in the period studied.

This then, is the challenge: an economy in which knowledge and skill needs are rising and in which, therefore, individual workers, teams and organizations must continuously acquire and apply knowledge in the pursuit of innovation.

3. TOWARDS THE LEARNING FIRM

The contextual features described in sections 1 and 2 are, as de la Mothe and Paquet (1997) put it, "forcing enterprises in the new economy to embody a philosophy of continuous improvement and innovation, to become learning organizations, and to do so, they require an organizational flexibility that they did not previously possess. This includes adopting a consultative coordination capability within the production chain, and a cooperation strategy with other stakeholders and governments".

Their exposition emphasizes certain aspects of organizational learning that are useful to bear in mind for the following discussion. They emphasize, for example, that the "new knowledge production" depends not only on cognition, but, critically, on interaction. Thus it calls for new modes of collegiality, and alliances for sharing of knowledge. The development of what the authors call "a consultative coordination capability" requires trust, social cohesion, and networks. Such considerations have enormous implications for policy since, in reality, much knowledge is tacit, segregated and distributed, its production is transdisciplinary, and the teams that produce it are heterogenous and diverse. So the challenge is daunting and the possibilities of failure are many.

3.1 Definitions

With these preliminary observations in mind, how has the learning organization (LO) been defined in the literature? First, while some writers distinguish between LO and organizational learning (OL) the distinction is perhaps a bit forced: for present purposes a learning organization is simply one that undertakes organizational learning. Following Tsang (1997), it is helpful to note that definitions tend to reflect the nature of the study in question --whether descriptive or prescriptive, for example-- and the author's perspective as to whether organizational learning results in cognitive change, behavioural change, or both.

Most definitions incorporate basic elements: cognition, change in actual behaviour, and change in potential behaviour. First, for learning to take place, knowledge, insight and/or understanding are gained (cognition); second, this is acted upon in such a way that improved methods, procedures, processes, policies or practices are introduced (change in actual behaviour); or, third, the learning process provides the basis, the launch-pad, for the introduction of the *new* policies, procedures and practices to meet *future* needs (change in potential behaviour) (Newton & Magun, 1998).

One of the more comprehensive definitions is offered by Senge (1990):

> When people continually expend their capacity to create the results they truly desire, where new and expansive patterns of thinking are nurtured, where collective aspiration is set free, and where people are continually learning how to learn together.

The level of learning is a further defining characteristic. While the individual is the basic unit for cognition, interaction is needed for knowledge to be communicated so the team and the organization are additional levels of analysis. Interfirm relationships, research consortia and firm-client partnerships constitute another level. Then one might envisage a "meta learning" level at which an economy's national system of innovation and learning infrastructure facilitate the acquisition of knowledge through networks of institutions and organizations. Finally, the phenomenon of globalization means that increasingly, firms' vision must transcend national boundaries and its learning must take into account the diversity of cultures, tastes, languages and ways of doing business.

3.2 Some Conceptual Considerations

Several recent approaches to the concept of knowledge are useful. Four categories of what they call "economically relevant" knowledge are distinguished in Lundvall and Johnson(1992). "Know-what" is factual, codified knowledge. "Know-why" involves understanding of the principles of, for example, science and society. "Know-who" is knowledge of who knows what and can do what --an important aspect of knowledge because of the growing importance of interaction, teams and coordination in the learning firm. "Know-how" is the tacit knowledge, the skill and experience, and intuitive "knack".

This latter form of knowledge might be exemplified by the canny, tacit, instinctive skills of a Pele in contrast to the codified diagrams of a coaching manual. A number of authors point to the transformation of tacit to codified knowledge (aided enormously by ICTs) as one of the central activities of the learning firm, though others point out that the transformation of new codified knowledge into new tacit knack and know-how is also a valuable form of learning. Still others (Foray & Lundvall, 1997) argue that the obsession with codification can be taken so far as to jeopardize flexibility and impede the learning process.

Nonaka (1991) in his classic article on the knowledge-creating company posits four patterns of interrelationship between tacit and explicit (or codified) knowledge. From tacit-to-tacit is dubbed *socialization* and is exemplified by the relationship between master and apprentice, in which the latter observes and mimicks her mentor. *Externalization* is from tacit to explicit: when innate skills, know-how and experience are systematically articulated and rendered into formally communicable form. Moving from explicit to explicit is what Nonaka calls *combination* --as when an engineer surveys a variety of designs, writes a synthesis report and recommends a new alternative. Finally, as suggested above, the movement from explicit to tacit (*internalization*) is also an important form of learning --e.g. when a new piece of codified information (a system, or software, perhaps) is used to the point that it becomes routinized and instinctive and the users discover or create their own short-cuts and tricks-of-the-trade.

In arguing that the KBE has ushered in not only a new cognitive division of labour, but also a new mode of production of knowledge, de la Mothe and Paquet (1997) following Gibbons (1994) and Gibbons et al. (1994) point to a mode of knowledge production that stands in sharp contrast to that of yesteryear. This new mode cuts across and combines various disciplinary approaches, topics and issues, may involve fleeting, flexible combinations of peers who produce knowledge that is more socially distributed and more socially accountable (in the sense of the multiplicity of stakeholders). This, they contend, is the world of what they call "Delta knowledge": knowledge that is the result of a creative, synergistic process of multidisciplinary learning-by-doing to address a particular issue or problem.

Finally, Leonard and Straus (1997) advance ideas about knowledge creation that have echoes of the emphasis on transdisciplinary teams and the need for consultative coordination capability. In describing what they label *creative abrasion* they argue that the successful innovating manager gets different approaches and ideas (not people) to grate against one another. In other words, get people to collaborate, and ideas to clash. Successful managers, they contend, are those who understand the value of deliberately bringing together people with different thinking styles: analytical or intuitive, conceptual or experiential, social or independent, logical or values-driven. They systematically design a full spectrum of approaches and perspectives into their organizations and simply establish the climate and the groundrules for the creative process. The resulting learning firm they refer to as the whole-brained (as opposed to right- or left-brained) company --one in which the knowledge creation stems from *cognitive diversity*.

3.3 Applications and Models of Organizational Learning and Intellectual Capital

The literature on applications of these concepts in practice, and the many attempts to capture the many and varied applications in model form, is voluminous. Some of the main works are synthesized in Newton and Magun (1998). For present purposes it suffices simply to mention some of the major elements that emerge.

In practically all cases there is an emphasis on the process of knowledge acquisition or production and its dissemination, utilization and application. As far as acquisition is concerned, some knowledge according to Huber (1991) is *congenital*: what the firm inherited or learned when initially "putting its act together"; some is *acquired* through systematic experimentation; some is *vicarious*: learned through benchmarking, identification of best practices, theft,[1] espionage, etc. Some is acquired through *grafting* (takeovers or mergers to gain new intellectual capital).

Another generic approach is to set out those characteristics of the firm that identify its particular approach to learning and then set out the various factors --policies, procedures-- that support and facilitate the overall approach. Thus the learning characteristics or orientation of the firm might include the following. Is this firm an innovator or an imitator? Is knowledge acquisition and dissemination facilitated by formal procedures or does it happen informally? Is the focus on incremental or transformative learning? Is knowledge acquisition an individual or group process? As for the supporting practices and procedures, these might include environmental scanning, the development and use of metrics for ongoing monitoring and evaluation, the encouragement of continuous learning and experimentation, and the leadership by committed and involved advocates and champions.

The profile of the concept of intellectual capital has acquired prominence somewhat more recently than organizational learning. But, of course, they are inextricably related: it may be logically subsumed under the latter: in a sense, intellectual capital may be thought of as the stock that is augmented by means (among others) of the process, the flow, of organizational learning. It consists of technical knowledge and skills; management skills; vision; experience, know-how, knack, flair and creativity; cognitive approach; and personal and psychological attributes including personality, ambition, energy, entrepreneurialism, etc.

Such has been the momentum of this concentration on knowledge assets that since the Scandinavian financial services giant Skandia first appointed Leif Edvinsson to the post of Director, Intellectual Capital, job titles such as Chief Knowledge Officer (CKO) have blossomed in several countries. However, despite the rhetoric, the practice of intellectual capital identification, measurement, development and management appears limited in scope (Newton and Magun, 1998).

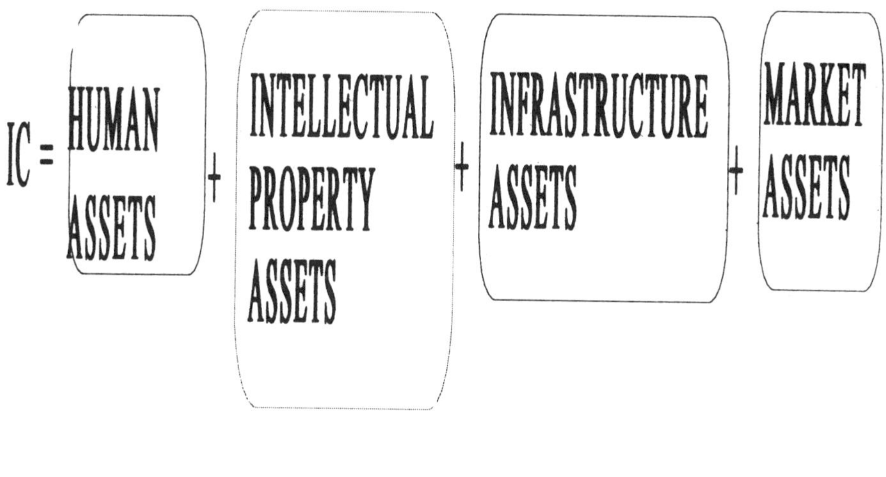

Figure 2. Intellectual Capital

A major challenge is conceptual. Even definitions are bothersome. Several influential writers, including Edvinsson and Malone (1997), Stewart (1997), and Canada's Hubert St. Onge (1996b) break IC into 3 major components: human capital; structural capital (information systems, policies, procedures); and relational, or customer capital, which consists of the firm's relationships with its clientele. However, our preferred definition is the more comprehensive one offered by Brooking (1997) and set out in the stylized equation depicted in Figure 2.

Human assets include individual skills, entrepreneurial and managerial competencies, and personal attributes such as leadership, determination, independence, helpfulness, extroversion or introversion, etc. Next, intellectual property assets would include patents, copyrights, licenses, and trade marks. Third, infrastructure assets include information and communication systems and software; organizational design; policies, practices and procedures; and corporate philosophy and management style. Finally, market assets include corporate image, reputation, brand loyalty and customer relations and satisfaction.

These, then, are the kinds of considerations that should be taken into account by firms that seriously and systematically seek to become learning organizations. They do so because of the perception that there is no alternative to the pursuit of knowledge if they are to continuously innovate. And continuous innovation is the route to success. Implicit in most of the models, therefore, is the notion that learning and the development of intellectual assets pay off in terms of enhanced competitiveness and various performance measures.

Accordingly, our generic, synthetic model in Figure 3 takes the principal components of the learning firm and suggests (since this is an empirical question) that they may lead to positive performance outcomes. Starting at the left are the main requisite elements for the learning firm which create the setting in which, in block two, organizational learning may occur at various levels. The process of learning creates the knowledge that is embedded in the various assets that constitute the firm's intellectual capital (in the third block). The fourth block emphasizes the need for knowledge management, while the fifth suggests some of the salutary effects that might ideally be expected to ensue.

Fig. 3: A General Framework For Analyzing the Elements of a Learning Organization and Intellectual Capital

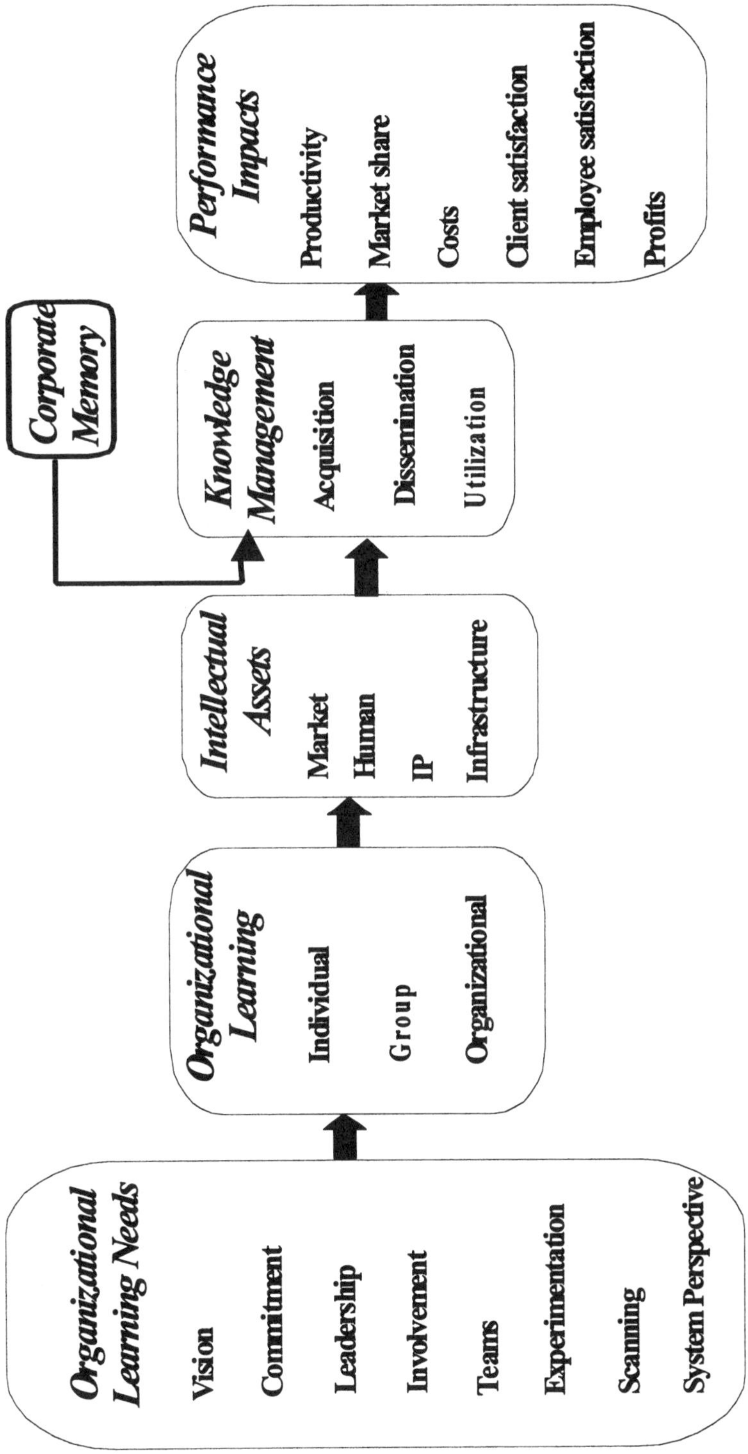

3.4 Empirical Evidence

Much of the voluminous literature on learning organization, and also on intellectual capital, comes from case studies and can be found in the pages of the management and business journals. More analytical research based on survey data is found in the literature pertaining to high-performance work systems (see, e.g., Gephart and Van Buren, 1996). Canadian empirical literature in this tradition is surveyed in Newton (1996,1998). Basically, that literature has two important messages that may usefully inform research and practice on the learning firm. First, there is a persuasive body of evidence showing that *bundles* of human resource practices, in concert with broad corporate goals and strategies, have had positive performance effects. This suggests that the elements of organizational learning set out in the generic model of Figure 3 should be mutually complementary. The second conclusion is also applicable to the learning firm and that is the importance of "intervening variables", that are critical in transforming policies, practices and procedures into outcomes. It may not be enough to simply have a policy in place. It is vitally important to provide the motivation, the involvement, the empowerment, the learning opportunities that lead to acceptance, commitment and the enthusiastic "buy in" that leads to enhanced performance.

There is a dearth of Canadian literature on the learning organization and intellectual capital *per se*. Exceptions are Stivers et al. (1996) and Richards and Goh (1995). The first is a comparative study of the management of intangible assets in US and Canadian firms. Senior executives were asked to indicate, on a 5-point scale, their level of agreement with statements about how their companies manage knowledge. Relatively high mean scores for both countries attest to the importance attached to intellectual assets. Scores were generally higher for the US on most questions, although there were few significant differences.

The second study is employee-based and poses 21 questions in five scaled clusters: clarity of mission and purpose; leadership and facilitation; experimentation and rewards; knowledge transfer; and team and group problem-solving. These are used to rank a federal government department, two special Operating Agencies, and two private sector telecom firms. A further description of these Canadian studies and some American work is contained in Newton and Magun (1998).

Some rather preliminary, though very promising Canadian work is a Study of Knowledge and Human Resource Management in SMEs in British Columbia. As reported in Schuetze (1999) the mail survey it utilized was the outcome of in-depth case studies in each of four sectors: computer services, engineering consulting services, value-added wood products manufacturing and telecom equipment manufacturing. Some interesting results include the following (in Schuetze's words):

- most knowledge to innovate comes from a network of customers, partners, suppliers, consultants, trade fairs and associations --not from universities or government labs.
- firms are notably less systematic when it comes to gathering *external* knowledge than they are at creating knowledge *inside* the firm
- most firms use traditional human resource management models
- firms talk the talk of HRD strategy but few walk the walk (no formal policies or operational arrangements)
- firms see the need to become "learning organizations" but marginalize human resource management
- separate training budgets and training are rare
- informal training arrangements greatly outnumber formal ones
- co-op and work-release programs are very popular with high-tech and engineering firms
- firms prefer to hire juniors, unless they need very specialized experience
- in general, firms are becoming less hierarchical and controlling, but the shift to a flexible workplace is far from complete
- management *style* is as important as management *structure*

Finally, some brief conclusions from eleven case studies in Ontario and Quebec (Newton and Magun, 1998). A questionnaire was used to structure the interviews for the case studies. Basically, in addition to the usual "tombstone" data and questions dealing with the firm's technological and organizational innovations, the bulk of the questionnaire focuses on motives for adopting OL, and the policies, practices, systems and facilities used to implement and develop it. Further questions dealt with the impact of OL on a variety of performance indicators and, finally, with firms' identification and measurement of IC and the extent to which they participate in joint research ventures and partnerships for knowledge acquisition.

The results showed few (though some) surprises. First, vision, commitment and leadership were crucial. Next, empowerment and teamwork were mutually complementary. A particular feature of the former was what might be called a "culture of personal responsibility" in which the firm commits resources and provides the setting in which individuals can self-assess and then take charge of their own skills and knowledge acquisition. As for teams, a common model was cross-functional groups, learning each other's perspectives, needs and goals, sharing knowledge, and spreading risk. This has echoes of the "cognitive diversity" referred to earlier and was often expressed in terms of "putting yourself in the other guys' shoes".

In all cases technology was important -- as a learning device, for communication, and for analysis. One somewhat unexpected result was that technology in addition to being a necessary condition for learning, is in some cases a motivator. "Techies", to put it crudely, are turned on by fast, powerful and sophisticated equipment and software (Newton and Magun, 1998).

The results also show instances of a systems perspective --i.e. the ability to envisage the various essential elements of the learning organization and the way in which they must interact in order to be mutually supportive. This was frequently expressed in terms of a "holistic view", or the "big picture".

A couple of organizations, one a privatized former government agency, and one a major IT company once poised on the brink of insolvency, had been forced to fundamentally rethink the basic underlying mission, goals, mandate, vision and strategy for their organization --what Argyris (1977) felicitously called "double-loop" learning-- not just responding to a problem, but rethinking the very fundamentals. As to models and metrics, some firms, especially in the financial sector, have clearly been influenced by contemporary management thinking and have taken steps to start assessing intangible assets and/or explore the possibility of applying some version of the "balanced scorecard" (Kaplan and Norton, 1996). In most cases there was some kind of external partnership for knowledge creation and transfer. Finally, as expected, the study found much enthusiasm for the concept of intellectual capital but with the exceptions mentioned above, few examples of concerted efforts to identify, much less measure, it.

4. MEASUREMENT ISSUES

In this section it is argued that while there are few existing surveys that deal directly with the features of the learning firm set out above, there are a number which afford the possibility of assembling a good deal of related evidence. There are, moreover, a number of questionnaires that can usefully be scrutinized to suggest the types of issues that should be considered. With the theoretical and conceptual issues of section 3, and the generic model of Figure 3 in mind, what are some of the key issues one might want to address at what might be called the macro/meso and micro levels, respectively? What do some of the surveys have to offer, and what are some of the gaps?

4.1 The Macro/Meso Level

Two of the richest sources of data pertaining to the learning firm derive from Statistics Canada. The first, the Survey of Innovation, elicits information on a large number of related topics including education and training, R & D, intellectual property, and the competitive environment. Of particular relevance are a number of questions relating to innovation. These include:

- new products and processes
- organizational innovations
- qualitative and quantitative impacts
- objectives of innovation
- sources of information for innovation
- particular technologies
- barriers to innovation
- expenditures
- importance of innovation
- cooperative ventures
- impacts on workers and their skill requirements

Second, the Workplace and Employee Survey also has a number of questions relating to technology usage, implementation and impacts, product and process innovation, work organization and organizational change, impediments, use of government programs, and a wide range of internal labour market issues.

One of the best-known Canadian longitudinal surveys is the Working With Technology Survey (WWTS) first developed at the Economic Council of Canada and subsequently re-run and analyzed by Ekos Research Associates and the Canadian Policy Research Network (CPRN). It has been conducted 3 times --first in 1985, covering the period 1980-85; next in 1991, covering the period 1986-91; and again in 1995, covering 1992-94. An establishment survey, it covers all regions and all industries except agriculture, fishing, construction and public administration. The first two waves of the survey collected information on the types of computer-based technologies (CBT) being put in place; employers' responses, especially with respect to training; and the extent to which organizational innovations had been put in place. The focus of the third wave shifted somewhat, to collect more detailed information on the nature of the skill impacts of CBT.

These surveys and others from Statistics Canada[2] furnish a great deal of what might be called "circumstantial evidence" about learning firms. One might argue, however, that more tightly focused instruments would yield a richer basis for analysis. In this context, one such instrument from which one might gain valuable insights is that used for the study reported in Schuetze (above). In addition to generic questions about business environment and strategy and an exhaustive section on training and HRD, it contains a number of questions relating very directly to the LO. These include issues such as the types of collaborative mechanisms that are used for knowledge creation; sources of information for innovative activity; workforce qualifications and traits; skill and knowledge needs; acquisition, analysis and dissemination of information; and the development of learning culture.

Next, though put forward as very much a preliminary and tentative effort, some of the issues in the questionnaire underlying the case studies for Newton and Magun (1998) might prove worthy of consideration. Lastly, the questions posed in the comparative study referred to above by Stivers et al. involve scaled responses relating specifically to perceptions of the importance of intellectual capital, such as: "knowledge is our company's most critical resource; intellectual capital is an integral focus of our performance measurement system; intangible resources are more important than tangible resources for gaining competitive advantage"; and others of similar ilk.

Taken together, these sources suggest directions in which future measurement and research might usefully go. In addition, however, we have, on reflection, some further recommendations about topics and issues that relate specifically to the learning firm. The first is that, in identifying and characterizing the learning firm, "climate" and "culture" are paramount. So one wants questions that establish whether there is vision, leadership and commitment. Is there a mission statement, strategic plan, written policies or procedures that relate to learning? How strongly do managers (and employees!) feel that the firm is committed to continuous learning?

Next, does the firm manifest some of the characteristics that one might associate with a knowledge culture? What proportion of the workforce have post-secondary education? What proportion are involved in R & D? Does the firm engage in research alliances or other partnerships for knowledge creation and sharing?

Are the policies, procedures and tools those that one might associate with a learning firm? Are there formal training plans and budgets, for example? Since, as indicated in section 3, interaction is crucial, one would also want to establish whether there are teams and mechanisms for information sharing and participative decision-making. What specific IT tools are used for information dissemination? (Some firms have on-line repositories of collective wisdom -- a-thousand-and-one practical problems encountered, and suggested solutions, to which employees are encouraged to contribute regularly.) Innovative HR practices would also be consistent with a learning firm: pay-for-knowledge; sabbaticals; educational leave; use of educational/training software -- "courseware"; suggestion plans. And in addition to statements such as "yes, we consider ourselves to be a learning firm" it would be useful to obtain information as to whether this is taken seriously in practice. So, for example, one might want to find out if the firm has used some form of diagnostic tool to assess itself in terms of learning firm criteria (of which more in the next subsection). Or one might want to ask whether the firm maintains a human resource inventory that records type and level of educational attainment, experience, skills and competencies, and psychometric attributes.

4.2 The Micro Level

In the last section we alluded to tools for firms' self-assessment of learning potential and intellectual assets. Referring once again to the generic model set out in Figure 3, a simple, but useful starting point might be to critically assess the firm in terms of the list of organizational learning needs at the left of the model. Under *vision* one would ask: does our company have the right vision and culture to reach its goals? What is the capacity of our environmental *scanning*, benchmarking, and identification of best practices? And so on.

Just a concerted discussion and assessment of these elements is a useful exercise in itself. But a more valuable approach involves three additional steps. First, the participants in the exercise collectively rate the firm on a scale, for each of the elements. They then (by a process of scanning) identify best practices, and finally they rate the firm against the best practices to estimate the gap.

As far as measuring the firm's intellectual capital is concerned, Brooking uses her definition to conduct firm audits. Working with senior managers she uses a pistol- or rifle-range target as a heuristic device. Concentric circles, starting with the "bull's eye" represent decreasing scale-scores. The target is divided into four quadrants representing her four categories of IC: human, intellectual property, infrastructure, and market assets, respectively. The company is then assessed on various dimensions of each asset, with maximum scores hitting the bull's eye.

One final note on assessment refers back to the notion of cognitive diversity alluded to in section 3 --that is, different people have different thinking styles and in some cases it may be effective to establish "creative abrasion". Accordingly, it is useful in some circumstances to identify how individuals think. One common way of doing this is by means of the Myers-Briggs Type Indicator, which assesses individuals in terms of 16 personality types. Somewhat less well known is the Herrman Brain Dominance Instrument (HBDI) described in Leonard and Straus (1997). The HBDI measures a person's preference both for right-brained or left-brained thinking and also for conceptual or experiential thinking:

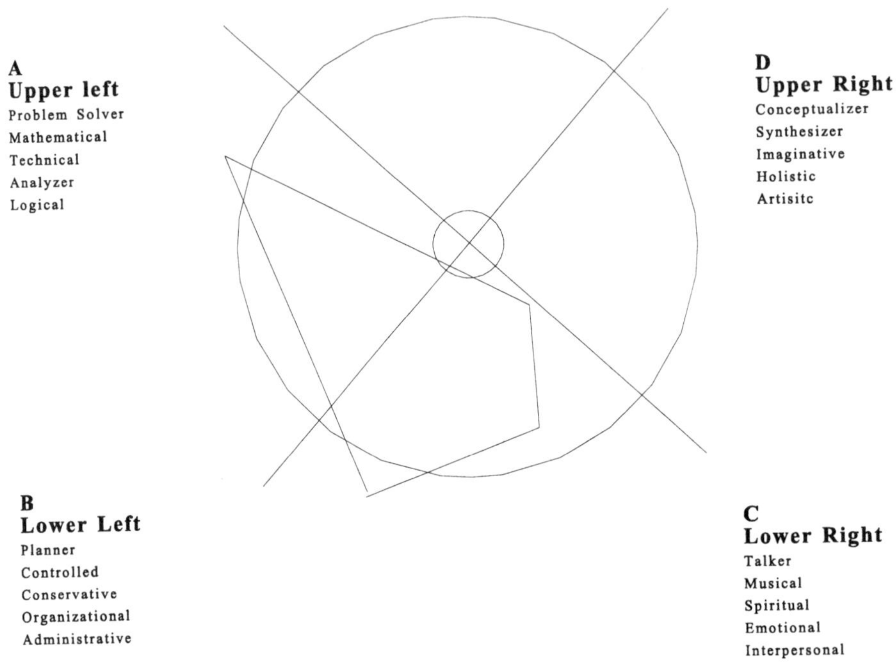

Figure 3. The HBDI

Source: Leonard and Strauss (1997)

Plotting individual members, one can determine the degree of homogeneity or heterogeneity of a team.

5. IMPLICATIONS FOR POLICY AND FURTHER RESEARCH

The foregoing discussion suggests policy implications for firms and for governments. At the level of the firm there are clear challenges to managers who intend seriously to address the issue of knowledge creation. The first, quite simply, may be how to change mind sets, since the literature clearly points to the importance of vision, culture, climate and commitment as prerequisites for organizational learning. Second is the business of addressing the "intervening variables" issue. Closely linked to changing mind sets, this refers to the somewhat more immediate day-to-day challenge of getting people to be sufficiently motivated as to actively and enthusiastically buy-in to the continuous learning process. This requires establishment of the appropriate incentives and rewards and may require especial innovativeness in the public sector context where there may be less flexibility in the conventional compensation system than is the case in the private sector

Perhaps the toughest challenge is minimizing the possibility of what de la Mothe and Paquet (1997) referred to as coordination and interaction failures, particularly in the light of the increasingly transdisciplinary and heterogeneous nature of the new knowledge production. And especially since, as Leonard and Straus (1997) suggest, there are instances where it is advisable to consciously establish cognitive diversity.

The question of coordination and interaction applies not only at the intrafirm level. Effective knowledge production requires the ability to work with other firms, individuals, universities, governments etc. This suggests a role for intermediaries who know enough about research to understand knowledge needs and who are also well versed in the skills of agency and brokerage. In some cases this might be a role for government. Indeed it is interesting that industrial policy has, for various reasons, moved away from the heavily interventionist role typified by subsidies and bailouts and increasingly toward government as the facilitator, the catalyst, the provider of strategic information. In addition, however, it is clear that research consortia not exclusively led by government --such as the Ottawa Carleton Research Institute (OCRI)-- have a valuable role to play.

The rationale for government facilitation in the area of organizational learning is three-fold. First, while there are powerful intuitive (and some empirical) reasons to suggest that such learning strategies pay off in terms of innovation and associated performance goals, there are concerns that firms are not adopting them as rapidly, as widely and as comprehensively as might be desired. The reasons may have to do with externalities: these types of strategies involve major investments in highly mobile human assets.

Another reason is imperfect information, since it is extremely difficult for (particularly small) firms to assess the costs and especially the benefits of systematic knowledge production strategies. In particular there is a dearth of solid empirical evidence to support the contention of positive bottom-line impacts. This latter point is important in connection with another possible rationale for government -- namely that there is a public good element to such learning strategies that is analogous to R&D and the adoption and diffusion of hard technologies. That is, if the strategies do pay off there is a collective benefit and the social rate of return may exceed the private.

Accordingly, what governments might usefully do would include the following. First, encourage the development of the infrastructure for knowledge production and dissemination: the research institutions, the electronic infrastructure, the networks, the information bases. Second, enhance its facilitation and brokerage role to develop partnerships and minimize coordination and interaction failures. Third, develop and diffuse diagnostic tools, especially for small businesses, and undertake or support research on: critical success factors; barriers; reasons for failure; and best practices.

Finally, as far as further research is concerned, here is an area sorely in need of a focussed project in which case studies might inform the development of a survey, preferably with longitudinal and international comparative dimensions. The design features might usefully include the types of questions raised in the measurement section (4a) above. Urgent research questions include the identification of learning firms: who are they, where are they, in what numbers, and what are their distinguishing characteristics? Second, what are the critical elements, and facilitating factors, that constitute the learning firm? Third, what is the relationship between organizational learning and intellectual capital development, on the one hand, and performance outcomes on the other? What are some of the critical impediments? What is the role for intermediaries in general, and government in particular? It is work of this kind that could prove invaluable in helping business entities move to assume the guise of the "learning firm".

Notes

[1] Garvin (1993) advises managers to invoke the SIS principle (steal ideas shamelessly).

[2] Such as the Survey of Operating and Financing Practices, used by Baldwin (1998) to develop indices of competencies in innovation, technology and human capital development.

Chapter 4

Organising for Information and Innovation

What do we know about organisational change in enterprises?

GRAHAM VICKERY
OECD

Recent analytical work suggests that there are large gains from more effective organisation and deployment of technology and human resources.[1] These gains can be summarised succinctly as follows:

> Output growth is highest in those manufacturing and services sectors that develop and use technology most intensively and have the highest skill requirements ...efficient interaction between manufacturing and business services is becoming a key feature of firm performance....[2]

1. MEASUREMENT

Despite the accumulating body of evidence suggesting that the context is changing radically for business and for value creation, and that substantial gains are to be made from new business strategies, economic theory and government statistics have not been able to keep up with the untidy reality of an emerging information-rich knowledge economy. Enterprises have changed business strategies, structure and behaviour and there is substantial anecdotal evidence of companies adopting strategies and tactics that range from management succession planning to more systematic management of intellectual capital of all kinds. These range from assessment of skills and competencies of workers to an exploding array of ICT applications that aim to improve communication and feedback across operating divisions and across hierarchical boundaries.[3]

The productivity and job gains associated with new technologies are best realised when firms make complementary investments in organisational change and upskilling. However, there is no static single model of firm organisation and firm strategy that automatically brings benefits. Firms adopt different strategies to improve performance in response to competitive pressures. These range from product innovation, improvements in quality and variety, customisation and upgrading of customer service, where employee skills and organisation are of key importance, to strategies based on price, product standardisation and varying the quantity of labour input. There is a continuum of firm strategies across these possibilities, with a variety of more or less flexible forms of organisation.

The focus in this chapter is on a set of strategies and organisational forms based on innovation, high skills, organisational flexibility and trust, often termed "high-performance work practices". These practices are typically based on employee commitment and less use of contingent (part-time, temporary) employees. However, in some settings individual firms have adopted mixed strategies, with a core group of high-skill employees and a quantitatively variable peripheral workforce. More generally, part-time and temporary working have increased in many OECD countries.

At aggregate level, enterprise and establishment surveys have been developed to identify more clearly the changes in business strategy, corporate structure, work organisation, and external relationships that enterprises are adopting. The purpose of these surveys is to provide a basis for evaluating the extent to which enterprises are creating "high performance workplaces", incorporating various elements of information and knowledge management, and the impact of such innovations on enterprise performance. They have been undertaken in a number of cases to improve the knowledge and understanding of the substantial changes taking place in the organisation of firms and the deployment of their assets and competencies, to enable governments to come to grips with the extensive restructuring taking place.

2. STRATEGIES

Two broad strategies can be observed: new forms of work organisation to better exploit technology; and more attention to investing in and managing intangibles assets -- particularly those linked to technology and human resources. These strategies have been characterised as being "high skill, high trust" work practices and work places.

A series of recent surveys shows that the flexibility associated with high-performance workplaces has positive impacts on firm and establishment performance, particularly in association with technology and more highly skilled workers (OECD, 1999). These large-scale cross-sectoral surveys of the characteristics and benefits of "high-skill, high-trust" work practices supplement and extend the plethora of case studies. High-performance workplaces are strongly associated with:

- higher labour productivity, better wage performance (due to the premium placed on skilled workers), and satisfactory unit cost performance due to enhanced productivity and improved quality of outputs, particularly when a range of organisational innovations are adopted (based on high skills, high levels of training, distributed responsibilities, innovative pay systems and, often, quality-based practices);
- higher sales, as better organised, more efficient firms create markets and capture market share, customer satisfaction is greater due to better product quality and improved customer relations, and the financial performance of firms adopting these organisational strategies is often better;
- positive employment performance (particularly when high-performance work practices are associated with technology adoption) in conjunction with higher labour productivity due to improved enterprise performance, and lower staff turnover due to better working conditions and higher wages.

Firms and establishments adopting new organisational structures have stronger and more productive external linkages with their customers and suppliers of inputs and services (OECD, 1996). Overall high-performance workplaces are based on two distinct sets of features: new ways of organising work to effectively exploit technology; and a greater premium placed on building and using intangible assets, most importantly technology and human resources.

However, the evidence shows that the ability to adopt new organisational structures, or adequately invest in intangibles, varies widely across firms, sectors and countries, with important effects on output, productivity and employment. While these issues are more or less common to OECD countries, there are considerable differences in terms of uptake of high-performance workplaces as well as in the policies that impact on them.

3. CHARACTERISTICS OF HIGH-PERFORMANCE WORKPLACES[4]

The high-performance workplace is a loosely defined "model" based on "high skill" and "high trust". Jobs are more complex, with more tasks and greater interdependence and communication among workers, firms, their customers and suppliers. Organisations are often simpler, as responsibilities shift to operators or autonomous work teams or are pushed out to suppliers. This is in marked contrast to earlier work organisation, based on simplifying tasks and jobs and organising them in complex hierarchies. There are many ways of organising work within these boundaries, and no simple "one size fits all" prescription of organisational attributes. They encompass such concepts as "total quality management", "quality circles", "continuous improvement", "autonomous team work", and participatory decision making combined with such human resource management practices as continuous training and innovative pay systems that reward ideas and skills.

Furthermore, it is the mutually reinforcing nature of these bundles of practices taken together that produces the positive results summarised above (Newton, 1996). Many of the stylised examples of new organisational forms were originally built on structures and strategies developed in the very different industrial and social contexts of Japan and Sweden, principally in automobile assembly and similar industries, in conditions of tight labour markets. Because these models appeared to provide a range of competitive advantages for their developers and adopters, the original models have been widely studied, adapted and adopted. Meanwhile, changing circumstances have led Japanese and Swedish firms to adopt some of the features of work organisation found in other countries. More than fixed ideas about single organisational "models", the key is flexibility and experimentation.

The broad idea underlying the new ways of organising work is that firms and organisations can achieve their objectives by using the innovative abilities of individuals more effectively, and that this is based on employing more highly skilled people and encouraging them to use their abilities more fully. To achieve these aims, minimum qualification requirements have risen, and greater emphasis has been put on continuous learning and training, coupled with stronger incentives for upgrading skills to improve performance. The new work organisation has some or all of the following features:

- marked specialisation of enterprises or business units (focus on "core" activities);
- horizontal inter-firm links for subcontracting (purchase of components or services that are part of the final product) or outsourcing (purchasing supporting business, professional and logistics services, and cleaning, cafeteria or other ancillary services), and the growth of inter-firm networking and clustering;
- effective use of technology;
- increasingly flattened hierarchies with greater importance given to horizontal communication and horizontal links, with less importance attached to vertical or hierarchical ones;

- information is gathered at more levels and channelled less hierarchically;
- authority to act is less dependent on hierarchical models of authority;
- employees are better trained and more responsive;
- multi-skilling and job rotation increase, blurring differences between work activities;
- small self-managing or autonomous work groups are common and take more responsibility.

Different combinations of these work organisation innovations tend to be found in about a quarter of all enterprises or establishments surveyed. Such innovations appear to be found in larger firms, those in more competitive product markets, those producing high technology and knowledge-intensive goods and services. Moreover, though it is difficult to prove causal impacts on company performance, there are strong associations between the above average use of various strategies and tactics associated with better information and knowledge management, and a variety of measures of company performance such as output growth, profitability, productivity, customer satisfaction, sales.[5]

These organisational features, their combination and importance, will vary according to the national setting, the sector and the size of firm or establishment. In the past, different forms of labour flexibility and adjustment in the workplace have been loosely grouped into two extremes – "functional" and "numerical" flexibility. Many of the characteristics of the new innovative forms of organisation described above are associated with "functional" flexibility, although they may be combined with elements of "numerical" flexibility. Furthermore, there are major differences between firm adjustment strategies adopted in countries that can be described as having "market-driven" approaches compared to those with more "consensual" ones. Some of the stylised features of these approaches are summarised in the following section.

4. DIFFERENCES ACROSS COUNTRIES IN FIRM-LEVEL ADAPTABILITY AND FLEXIBILITY

Two different kinds of flexibility are described in the literature on strategies for using labour at firm level. Functional flexibility involves high-skill, collaborative approaches to work based on high quality labour inputs, often made up of a "core" group of long-tenured, educated and trained employees. Common features are shifting job design and job boundaries away from traditional narrow ones, mobility across tasks, multi-skilling and wide-skilling, extensive training and retraining. Autonomous self-managed multi-functional teamwork is often associated with this kind of flexibility. Numerical flexibility usually involves changing quantities of labour input. These include numbers of employees, hours of work, use of more peripheral part-time and temporary employees, and making use of liberal provisions on hiring and dismissals, usually in countries with lower hiring and firing costs.

There is also another dimension to adjustment. Internal flexibility is within the enterprise or the existing contract structure of the enterprise. External flexibility involves interaction in markets outside the firm, usually requiring changes in the nature and type of contracts. There is much overlap between internal flexibility and functional flexibility, and external flexibility and numerical flexibility. The country groupings that follow correspond to different patterns of institutional and policy arrangements that influence the way firms adjust and adopt different kinds of flexibility. These patterns of firm-level adjustment in turn are the determinants of wider patterns of adjustment at sector and country levels.

Consensual or relations-based approaches: Firms in Austria, Belgium, France, Germany, the Netherlands, the Nordic countries and Switzerland have adopted functional/internal strategies, based on long-term, skill-based contracts between enterprises and workers, with education and training systems often investing heavily in building a deep skill base and high levels of individual competence. This can be described as an economy-wide "consensual" (or "relations-based") approach, typified by extensive negotiation among a broad range of stakeholders to reach consensus, with restricted capital markets, and concentration and cross-holdings of capital ownership in banks and corporations. Collective bargaining is conducted through consultation at the broad industry level.

Firm-based consensual approaches: Japanese firms have also adopted functional/internal forms of flexibility, but adjustment is more centred at the firm level. As in Europe, restricted capital markets and concentration of capital ownership have often led firms to focus to a greater extent on achieving objectives such as market share and technology development rather than short-term financial performance. Labour adaptability is achieved through broadly-based general education and highly developed human resource development and workplace training strategies within enterprises to achieve flexibility and underpin enterprise competitiveness. Workplace negotiations are based to a large extent on firm-level arrangements, and overall it may be described as a firm-based "consensual" approach.

In contrast, in countries where firm strategies can be described as more "market-driven" (Australia, Canada, New Zealand, the United Kingdom, the United States), firms have adopted numerical/external strategies to a greater extent, based on use of external markets to adjust to change. Influenced by the relatively liberal structure of factor and/or product markets, and an economic environment that encourages entrepreneurship and risk-taking, firms have followed a variety of strategies to achieve adaptability and manage risk, while maximising shareholder value and satisfying other short-term financial criteria. Flexibility has been based on enterprises being able to add and shed fixed assets through take-over and divestment strategies in well-developed capital markets, and have recourse to well-developed labour markets external to firms to hire workers with required skills and dismiss those whose skills and qualifications are no longer needed.

More heterogeneous approaches to firm-level adaptability and flexibility can be distinguished in "intermediate" or "catch-up" countries (Greece, Ireland, Italy, Portugal, Spain, Turkey). Firms are usually less technologically advanced and fewer have adopted new models of organisation, although foreign multinationals often use advanced organisational strategies. Firms operate in more traditional institutional settings, often more rigid than in the "market-driven" approach. These settings may steer firms to adjust labour supply internally, even though firm adjustment is constrained by poor worker qualifications. Participation in regular education is below the OECD average, firm training effort low. These countries generally have average or below average rates of union membership, but high levels of collective bargaining coverage. Regarding the group of countries which have more recently joined the OECD grouping, firm-level adjustment strategies are still evolving. Firms in part to adjust as in neighbouring countries: the Czech Republic, Hungary and Poland similarly to northern European adjustment strategies; Korean firms along the lines of firm-level "consensual" approaches; and Mexican firms with elements of strategies in Canada and the United States. But recent institutional changes, economic upheaval and new laws have rapidly altered the institutional environment for firm-level adjustment, or introduced entirely new elements.

There have been recent shifts in the broad patterns described above. Countries that have typically greater recourse to external markets and strategies based on numerical flexibility have shown greater interest in making more functional and internal adjustments within the firm. Interest in the "high-performance enterprise" and in best practice in countries such as Australia, Canada, the United Kingdom and the United States is largely about how to build functional flexibility and improve the quality of labour and other inputs. In the European setting, the shift has been towards greater use of numerical flexibility and external functional adjustments as product, capital and labour markets are liberalised. Examples are the increases in Italian self-employed, growing levels of short-term employment in countries such as France, signs of increased outsourcing in German manufacturing, and the growth in temporary job agencies wherever they are allowed. This has paralleled major restructuring efforts in many firms and industries, as greater competition, initially from Japanese firms and subsequently from North American ones, has led European firms to rapidly adopt quality and efficiency practices such as "just-in-time", "total quality management", and the like to improve competitiveness (see OECD 1996 and 1998).

5. ORGANISING FOR INNOVATION

There are also differences across countries in the organisation of innovation (OECD, 1992). The United States' approaches were characterised as extensions of the "Taylorist" organisation of the factory floor. This stressed deep, but narrow, technical specialisation, separation of functions and local functional responsibility. In large US and European automobile firms, a separate research organisation interacted with product development, with minimal interaction with manufacturing and distribution. In contrast, large Japanese manufacturing enterprises appeared to integrate and overlap research with all phases of product development, manufacturing and distribution. Innovative ideas were encouraged from customers and employees, with major aims being design for manufacturability and minimum time to reach the market place. The overlapping approach required effective information transfer, feedback and dialogue, using cross-functional task forces.

Under pressure from efficient Japanese industrial innovation techniques, large firms in the United States, followed by European firms, have increased efficiency of central corporate activities, including technology functions. This has been part of a strategy to become more market-responsive and maximise shareholder value. R&D is organised on a centre-of-excellence model, and human resource implementation is decentralised. The optimum size of central functions is often seen as several hundred people (Boston Consulting Group, 1996). Team-based approaches have proliferated. Detailed surveys in Germany show that R&D and creative management tasks are often organised in high-level autonomous work teams, particularly in technical and business services and other producer services (Kleinschmidt and Pekruhl, 1994). This suggests that approaches seen to give Japanese organisation the edge were adopted widely in R&D and technology areas, at least in other "consensual-based" countries such as Germany. In the meantime, Japanese organisation of innovation has also been undergoing change. In the 1990s, cost reductions have come increasingly from lean product design and simplification of incremental innovations (Fujimoto, 1998).

Turning specifically to SMEs, there is growing evidence that innovative small firms have many of the organisational characteristics of large firms, emphasising human resource development and developing intensive networks with other firms. In contrast to large firms, however, they rely on customers and market information rather than R&D departments for innovative ideas, and speed to market is critical for protecting intellectual property; relatively little attention is paid to patents (OECD, 1996; OECD, 1997). Organisational barriers to innovation vary markedly. In countries such as the United Kingdom and the United States, there is excess demand for skilled labour. Firms in northern European countries appear to have greatest problems translating technological advance into marketable products and entering non-traditional markets. "Catch-up" countries show weaknesses in the scientific, technological, financial and business services infrastructure which inhibit innovation and productivity growth. These patterns reflect broad differences across countries in entrepreneurship, firm start-up and growth and more generally in their labour and capital markets. Ireland, the United Kingdom and the United States generate many innovative new and small firms which experience turbulence in surviving. In contrast, start-up rates and growth of firms are lower in some northern European countries (*e.g.* Finland and Sweden, but not Germany) and Japan.

There is no single best way to organise innovation, either for large or small firms. Flat organisational structures and streamlined business processes improve efficiency in delivering products, but other flexible approaches are also evolving (Tidd *et al.*, 1997). Organisation and the efficiencies derived from it will differ across countries, sectors and firms. The broad approach to policy is to improve education and the preconditions for skill development and, where necessary, remove barriers or provide stronger incentives for experimentation and adoption of new ways of organising innovation. This may be achieved for example, by improving mobility of staff working on R&D and innovation and ensuring that the government science system has strong but flexible links with business.

6. THE INCIDENCE OF HIGH-PERFORMANCE WORKPLACES

By the mid-1990s, it is estimated that the high-performance workplace model had been adopted by about a quarter of all enterprises in the OECD area. The level of adoption appears fairly similar across countries, suggesting common competitive conditions and similarities of practices. An increasing number of countries are carrying out large cross-sectoral surveys of various aspects of organisational change; but coverage is generally better in manufacturing than for services, and definitions of the new work organisation vary (Vickery and Wurzburg, 1998). Adoption appears to be somewhat higher in the United States and northern European countries – up to one-half of responding establishments or firms. However, the responses depend partly on the kind of organisational changes being surveyed, and whether change is defined as using a range of different techniques to suit firm circumstances, or whether stand-alone individual characteristics of organisational change are considered (OECD, 1996).

The only multi-country cross-sector survey (of direct employee participation in organisational change) shows that firms in northern Europe were more likely to adopt bundles of initiatives for organisational change. Workplaces in industry, construction and services in France, Italy, Portugal and Spain were less likely to be involved. On the other hand, workplaces in Denmark, Germany, the Netherlands and the United Kingdom have high levels of organisational change initiatives, with or without direct employee participation. Ireland and Sweden fell between these two groups. Thus there is a distinct north-south divide, with extensive workplace change in northern Europe, less in the south (European Foundation for the Improvement of Living and Working Conditions, 1997). In this survey, the principle motives given for organisational change were to improve productivity and performance, and the prime force driving them was increased competition.

Most of the surveys show that new kinds of organisation have been widely adopted in manufacturing. Assembly industries, notably automobile producers, have been much-studied and are often thought to typify the new kinds of organisation, emphasising quality and flexibility, reducing capital use, and shifting from vertical integration to horizontal supply arrangements, with external suppliers increasingly responsible for development of components. In services, change has been most pronounced in financial services and other tradeable and business services facing mounting competition. However, the extent of change by sector depends on what is being measured. The European survey found manufacturing to be lagging public and private services in the use of direct participation – one indicator of organisational change (European Foundation for the Improvement of Living and Working Conditions, 1997).

Finally, size counts: large firms and multinational firms are more likely to adopt new organisational forms than smaller ones. Small firms may in fact not need to adopt formal kinds of organisational innovations, simply because their more flexible structure means that these are *de facto* part of their operations. However most surveys show that large and medium-sized firms are more capable of adopting new flexible forms of organisation and may also draw greater productivity gains and other benefits from adoption (OECD, 1996; Lund and Gjerding, 1996; Ministry of Labour, 1996).

7. DOWN-SIDE THREATS

Despite these evident advantages of adopting new forward-looking strategies, there is little evidence of consistent corporate approaches to improving management of information and knowledge. A 1997 Knowledge Management Survey of respondents claiming to work for "knowledge-intensive" organisations (92 percent) showed:

- Only 6 percent of respondents described their organisations as being "very effective" in using and multiplying knowledge to improve business performance.
- Only 12 percent of respondents reported their organisations as being extremely good or very good at facilitating knowledge growth through enterprise culture and incentives.
- Only 9 percent were extremely good or very good at embedding new knowledge in the organisation.
- Only 3 percent were extremely good or very good at measuring the value of knowledge assets.

A later survey of a broader cross-section of practitioners found nearly 90 percent reporting costly mistakes for those reasons (Chase 1997). Slow progress in making knowledge management operational at the level of the enterprise translates into uneven transition to the knowledge economy. The slow pace stems in substantial part from the substantive complexity of the ideas embodied in "knowledge/information management", unclear definitions, and limitations on measurement. These in turn feed a confusing competition of ideas about how far knowledge can be defined, measured, managed, and owned.[6]

Quite apart from what appear to be positive effects of "high trust, high skill" strategies on performance, the reverse also appears to be true. Firm level performance may not be as good in a downsizing and unstable environment. New work becoming available suggests that organisational downsizing may yield short-term benefits for firms, it may undermine their capacity to innovate and secure their longer-term future. For example, organisational downsizing appears to have a negative effect on R&D workers' willingness to take risks, make suggestions and be motivated to innovate as opposed to being fearful of the work environment (Bommer and Lalajas 1999). Such threats appear to have a significant effect on the willingness of R&D workers to be innovative and hence improve long-term firm prospects.

8. POLICY CONSIDERATIONS

The broader impacts of firm-level organisational and technological change and adoption of the high-performance workplace, on employment, for example, depend not only on firm size and sector, but most importantly on the national institutional and economic setting. Some of the policy approaches to workplace adjustment aiming at improving enterprise efficiency and contributing to overall growth, productivity improvement and employment, are discussed below.

From the point of view of governments, there are important issues that arise in the midst of all this, issues that affect everything from levels of unemployment, levels and distribution of income, productivity growth, standards of living, competition and national competitiveness. Whatever governments do, there is a strong political consensus in favour of ensuring that government action rely as much as practical on market forces for achieving its goals. If governments start with the assumption that a knowledge/information economy builds "from the ground up", from the level of enterprise where factors of production are put together to produce goods and services, what role can they play in facilitating the transition to the knowledge/information economy?

9. ENCOURAGING FLEXIBILITY AND ORGANISATIONAL CHANGE

A number of impediments to the diffusion of organisational change limit the effectiveness of change. The nature and seriousness of these institutional barriers vary across countries. Notions of "systemic" approaches to policy depend on the broad set of national institutions, infrastructures and incentives which are explicitly or implicitly in place.

Approaches to adaptability and flexibility can be grouped into: *(i)* a "market-driven" approach in the English-speaking group of countries – Australia, Canada, New Zealand, the United Kingdom and the United States; *(ii)* a more "consensual" or relations-based approach in northern Europe; *(iii)* a "firm-based consensual" approach in the Japanese model; and *(iv)* a heterogeneous set of approaches in "intermediate" or "catch-up" countries (Box 1). This broad categorisation is useful to simplify discussion of policy design and country-specific policy approaches. Policy responses broadly need to combine the experimentation and flexibility that are part of the numerical/external strategies common in "market-driven" settings, with the longer-term skill formation and abilities to diffuse and adopt innovations that are part of firm functional/internal strategies common in "consensual" settings.

Policies to improve the adoption and impacts of high-performance workplaces are discussed below in five broad groups: *(i)* encouraging innovation; *(ii)* accelerating diffusion of organisational innovations; *(iii)* raising skills; *(iv)* encouraging labour-related flexibility; and *(v)* co-ordinating and delivering policy. Different aspects of the policy framework and policy initiatives fostering adoption of high-performance work practices are shown in Table 1. These show relative national strengths, corresponding to: policies to provide information on organisational strategies (usually but not always focused on SMEs); initiatives to improve skill formation (incentives/levies to encourage firm-based training, and vocational training); indicators of incentives to improve employee performance (profit-sharing); and indicators of flexible working practices and part-time employment as measures of labour flexibility.

10. ENHANCING INNOVATION

The development and adoption of high-performance workplaces and practices are examples of social and organisational innovation. Policy strengths and weaknesses across countries can be grouped to broadly follow the adjustment patterns outlined in Section 4:

– Experimentation and innovation with new work practices may be easier in countries where business regulations are relatively straightforward, and there is liberal product and labour market regulation. The ease with which firms can start up is an indicator of a pro-entrepreneurial approach which encourages business experimentation, including with work organisation. The United States is usually seen as leading in these areas, and there has been a great deal of policy interest in emulating some aspects of its framework. The United Kingdom and an increasing number of continental European countries have very respectable rates of new firm formation. On the other hand, there are major weaknesses in start-ups in some Nordic countries (Finland, Sweden), and also in Japan.

Table 1. Fostering high-performance workplaces: current policy practice

	Consultancy / information for SMEs	Incentives for firm-based training	Vocational training	Incentives for employee effort[1]	Flexible work	Collective bargaining / level	Co-ordinated policy approach[2]
Canada	•		Reform efforts	•	Liberalised/ part-time work common	Enterprise/plant	•
United States	•		Reform efforts	•	Liberalised/ part-time work common	Enterprise/plant	
Australia	•		Reform efforts		Liberalised/ part-time work common	Enterprise/plant & sector	
New Zealand			Reform efforts		Liberalised/ part-time work common	Enterprise/plant	
United Kingdom	•		Reform efforts	•	Liberalised/ part-time work common	Enterprise/plant & sector	•
Austria	•		■ Reform efforts			Centralised	
Belgium	•	•	■			Centralised	
France	•	•	■ Reform efforts	•	Liberalised	Sector	
Germany	•		■	•		Sector	
Netherlands	•	•	■ Reform efforts	•	Part-time work common	Enterprise/plant & sector	
Switzerland	•		■ Reform efforts		Part-time work common	Sector	

Denmark	•	•	■		Part-time work common	Sector	•
Finland	•		■ Reform efforts		Liberalised	Centralised	•
Iceland	•		■ Reform efforts		Part-time work common		
Norway	•		■ Reform efforts		Part-time work common	Centralised	•
Sweden	•		■		Part-time work common	Sector	
Japan	•		Reform efforts	•	Part-time work common	Enterprise/plant	
Italy		•	Reform efforts	•	Liberalised	Sector	
Ireland	•	•	Reform efforts		Liberalised		•
Mexico				•	Part-time work common		
Portugal	•	•	Reform efforts			Sector	
Spain	•	•	Reform efforts		Liberalised	Sector	
Turkey					Part-time work common		
Hungary		•					

- Ease of financing new ideas may also contribute to experimentation and adoption of new ways of organising work. Capital markets to finance new ideas have worked much better in the United States than elsewhere. In northern Europe and Japan, more relations-based investment has favoured established enterprises over new ones. In Japan and Korea, rather than searching more widely for new firms to invest in, banks have tended to retain investment within narrow conglomerate groups, encouraging incremental innovation but hindering more radical breakthrough.

11. ACCELERATING DIFFUSION OF ORGANISATIONAL INNOVATIONS

Despite growing evidence of the importance of organisational change and the impact of "high-skill, high-trust" work practices on firm productivity and performance, policies directly aimed at diffusing firm-level best practice are not widespread for a variety of reasons. It is difficult to define the characteristics of these work practices and strategies, responsibility for implementing change rests with employers and workers, and most countries lack a broad policy approach to building new frameworks for business. Furthermore, human resource issues are not typically part of industry policies, while education and training and labour market policies have tended to ignore firm and industry dynamics, focusing on formal compulsory education, and external labour markets and unemployment. However, shifts in the economic environment are challenging firm structure and performance, and there is a potentially important role for policies which enhance the transition to new conditions. The growing importance of small firms and their low levels of training and investment in intangibles need to be taken into account. Finally, policies need to focus on human resources and human capital accumulation for long-term economic development.

Government intervention may be justified if there are market failures in the adoption of high-skill and high-trust strategies and structures. Firm performance could be improved by better information flows on organisational change, management strategies, quality improvement, "benchmarking", through, for example, demonstration activities, identification and exchange of information on best practices, or activities which lead to continuous improvement and upgrading at the firm level.

Policies to diffuse best practice in organisational innovation and human resource development often consist in a broadening and re-orientation of established policies for technology diffusion. They have major differences, however. The focus is no longer on technology push, but rather on implementing systems of continuous improvement in firms across the whole range of firm functions, including organisation, quality, marketing internal and external networks, but most importantly including much more attention to skills and encouraging employee initiative. This places great demand on delivery mechanisms (consultancy quality assurance is increasingly important), and better articulation with education and training programmes. To be effective, programmes to promote organisational change and upskilling need to: have a broad, but manageable, scope, be correctly targeted, raise awareness, ensure programme uptake and have appropriate programme flexibility. Delivery is usually more efficient if it uses existing "demand-driven" institutions and infrastructures.

Adoption of high-performance work practices: In "market-driven" countries, the focus on short-term productivity and profitability may undermine long-term investment in human resources, organisation and technology, and the diffusion infrastructure may equally not be adapted to fostering long-term continuous improvements in firms. "Consensual" countries benefit from stronger infrastructure of supporting institutions, intermediaries, industry associations, unions, etc., all of which support diffusion and effective adoption. However, there are rigidities in some "consensual" countries, and traditional institutions providing business support need to become more demand-driven and pay more attention to new issues and challenges (*e.g.* the German experience with the long-standing and moderately well-funded Work and Technology R&D Programme shows difficulties adapting to change and in disseminating results; see Fricke, 1997). Austria, Canada, Iceland, Ireland and Norway have all introduced initiatives to address some of these issues. In the group of intermediate "catch-up" countries, greater effort must go to ensuring that incentives are right for firms to build internal intangible assets and adopt flexible organisational structures, and that an efficient demand-driven, customer-focused business infrastructure provides appropriate information and services to firms.

12. RAISING SKILLS TO ENHANCE ADOPTION AND IMPACTS OF HIGH-PERFORMANCE WORKPLACES

There is abundant evidence that past national strategies for investing in education and training have paid off in terms of faster productivity growth and higher levels of productivity at the aggregate level, and higher earnings and employability at the individual level (OECD, 1997; 1998). Firm-level high-performance workplaces and work practices depend on "high-skill" strategies that make better use of and continuously renew human capital. This suggests two essential elements in national strategies to reinforce wider adoption of high-performance work practices: *(i)* provide the broad infrastructure for compulsory and basic education, improve the quality of education, and encourage a larger proportion of people to obtain post-secondary degrees, including improved technical and vocational qualifications; and *(ii)* improve training and retraining of workers, and strengthen incentives for employers and employees to undertake training. These can be summed up in the concept of "lifelong learning".

Firm-level high-performance work places and work practices depend on "high-skill" strategies that make better use of, and continuously renew, human capital. But most countries need to strengthen incentives for employers and employees to undertake work-related training. Few direct incentives exist for firms to undertake training apart from tax-deductibility of training costs. Training levies are relatively rare. Tax incentives to individuals are even less common, with incentives weak in many countries, although some have been trying to strengthen such incentives.

In summary, broad directions for education and training policy reform arc as follows: improving the broad education and training infrastructure and making it more accessible, and improving work-related training, and strengthening the incentives for employers and employees to train.

13. ENCOURAGING LABOUR-RELATED FLEXIBILITY

To enhance efficiency firms adopt strategies ranging from "high-skill" and "high-trust" organisation through to changing the quantities of labour used. It is suggested that: *(i)* high-performance work practices will be more productive if they involve innovative pay systems such as pay for performance, profit-sharing or bonuses; and *(ii)* flexibility is enhanced and experimentation encouraged where fewer restrictions exist on working arrangements, but where the benefits associated with standard working (particularly training and skill development) apply also to non-standard arrangements, prorated as appropriate.

Policy directions to encourage greater labour-related flexibility, which will enhance the spread and impacts of high-performance work practices, can be summarised as follows:

- Performance-related compensation appears to be increasing across OECD countries. This is often associated with high-performance workplaces, productivity gains and improved performance. There is a case for expanding information on the benefits to be drawn from such practices. Finland, France, Mexico, the Netherlands and the United Kingdom have moved in this direction over the last decade or so; other countries, including Scandinavian ones, should consider following suit and modifying legislation where it prevents the spread of such practices.

- Flexible working hours and greater variability in work time (such as annualisation of hours) are probably useful adjuncts to the spread of high-performance work practices, as they allow enterprises to more closely align activity with demand. Recent initiatives have been taken in Finland, France, Greece, Italy and Spain (see OECD, 1997).
- Part-time employment increased in most OECD countries in the 1990s. High-skill, high-trust strategies based on skill formation and security may be undermined to the extent that such employees do not receive benefits and training. To promote new forms of flexibility, countries are generally reducing thresholds in hours worked below which employees cannot receive benefits, although Germany and Japan maintain relatively high thresholds. To the extent that thresholds inhibit skill formation they should be reviewed and eliminated or other targeted policy action taken to overcome them.
- Evidence for achieving organisational flexibility is apparent in the growth of "agency work" (typically a small proportion of total employment), most notably in the United States but increasing in other countries. The development of a competitive agency-work industry could help to provide flexibility to employers and more stability to temporary employees. Agency work and standard employment need to be subject to similar rules. Recent initiatives in the Netherlands are a good example, providing employers with labour force flexibility and employees with necessary security.
- Evidence on part-time and temporary work highlights "marginalisation" of younger and older workers. Apart from major issues of social cohesion, to the extent that these developments undermine more widespread adoption of productivity-enhancing work practices, they require attention, including, for example, generating greater "equality" between standard and non-standard forms of employment, and increasing coverage and portability of, *e.g.* pensions and health benefits, and improving access to training across groups of individuals.

Overall, high-performance work practices will be more productive if they involve innovative pay systems such as pay for performance, profit sharing and bonuses. Flexibility is enhanced and experimentation is encouraged where fewer restrictions exist on working arrangements, but where benefits associated with standard work, (particularly training and skill development) apply to non-standard arrangements, pro-rated as appropriate.

Pay for individual performance and pay for group performance (e.g. share options) are likely to increase, continuing to undermine traditional notions of wage setting and industrial relations.

14. FURTHER POLICY AREAS REQUIRING ATTENTION

Although they are not the prime focus of this chapter, there are other crucial areas requiring attention if the impacts of organisational change and the benefits of organising for innovation are to be achieved.

- *Accounting:* Periodic backward-looking financial reporting is becoming increasingly irrelevant as a guide to future performance of firms and economies. Better ways of coming to grips with information, knowledge, human capital and organisation are needed and are being developed through experimentation.
- *Intellectual property rights:* Information, knowledge, competence and related assets, their exploitation, the development of new products and services and new enterprise formation are the driving forces for competitive advantage, making intellectual property rights even more critical than in the past. Intellectual property rights need constant scrutiny to ensure that they encourage wealth creation and reward invention and innovation while improving diffusion of information and knowledge.
- *Competition policy:* New forms of internal and external business organisation, particularly in information and knowledge-intensive areas are likely to shift the boundaries of firms in radical new ways. This is changing the nature of competition as business models change and major segments of the value-added chain are re-shaped.
- *Taxation:* Information and knowledge are mobile, and cross-border transactions are growing rapidly. Revenue collection associated with exploitation of information and knowledge is likely to become more complex along the value chain.

15. CO-ORDINATING AND DELIVERING POLICY

Key ingredients in policy to improve the spread and impacts of high-performance work practices and foster investments in intellectual capital and intangibles are: encouraging innovation and entrepreneurship; improving information flows and demonstrating the benefits of high-performance work practices; ensuring that education and training systems and incentives provide a large pool of skilled employees and that firms undertake appropriate training; and ensuring that labour market regulations do not unnecessarily hinder the adoption of flexible strategies, and that attention is paid to access to benefits such as training and employee security, and that social cohesion issues are taken into account.

These need to be implemented so that they enhance organisational experimentation and market-led adoption by firms and employees. The agenda cuts across departmental and ministerial responsibilities in most countries. Co-ordinated and integrated approaches are likely to be most efficient (in the allocation and use of government resources) and effective (in enhancing firm uptake of high-performance work practices). The areas of responsibility that need to be involved in co-ordination cover science, research, industry, education and labour markets. Policy must be planned coherently, involving dialogue and co-operation with other actors (industry associations, labour representatives and other institutions), delivered at the appropriate level to reach targets (for example, SMEs), and monitored and evaluated to ensure that it meets criteria such as accountability, value for money and the achievement of objectives. The elements essential for policy design and delivery can be found in approaches adopted by a few countries.

Common threads run through approaches adopted.

- recognising the complexity of workplace change and adopting co-ordinated approaches;
- focusing on important elements in the national setting (information, demonstration, skills);
- involving business and labour organisations to ensure that initiatives are demand-led;
- working with specialised delivery agencies to ensure effective contact with target groups;
- choosing appropriate levels at which to co-ordinate policies and at which to deliver programmes.

Notes

1 This paper draws on work prepared by the author for OECD (1996), *Technology, Productivity and Job Creation,* Vol. 2, Chapter 6, "Technology and the changing nature of jobs: organisational change and human resources"; and OECD (1998), *Technology, Productivity and Job Creation, Best Practice Policies,* "High-performance workplaces and intangible investment". The views expressed here are those of the author and do not commit the OECD or its member countries.

2 OECD (1998), *Meeting of the Industry Committee at Ministerial Level: Scoreboard of Indicators*, Paris.

3 For case studies and models see Gallagher (1997), and Skyrme (1998). For a set of enterprise-level case studies covering a wide range of OECD countries see: http://www.oecd.org/dsti/sti/industry/indcomp/act/Ams-conf/symposium.htm. For general OECD work on intangibles see:http://www.oecd.org/dsti/sti/industry/indcomp/prod/intang.htm.

4 Work on high-performance workplaces has been undertaken in OECD projects on Technological and Organisational Change and Labour Demand/Flexible Enterprise: Human Resource Implications. This work was reviewed at an International Conference, Ottawa, December 1996, and published in Government of Canada and OECD (1997). Work on intangible investment is being undertaken in a cross-cutting project which resulted in an International Symposium, Amsterdam June 1999, *Measuring and reporting intellectual capital: Experience, issues and prospects,* results and papers available on the web-site: http://www.oecd.org/dsti/sti/industry/indcomp/act/Ams-conf/ symposium. htm

5 See Bassi and Van Buren (1998); and OECD (1997), *Lifelong Learning to Maintain Employability,* (OCDE/GD(97)162), Paris.

6 See also *California Management Review,* Special issue on Knowledge and the Firm, Spring 1998.

Chapter 5

Capturing the Elusive Social Impacts of Technology

Towards a Research Agenda[1]

SUSAN A. MCDANIEL
University of Alberta

1. INTRODUCTION

Canada, like many other industrial countries at the millennium, seems, as John Meisel so articulates it, "hell-bent on pursuing, through S&T [Science and Technology], the flickering grail of wealth, economic competitiveness, and something ambiguously called 'the quality of life,' defined invariably in terms of purely physical well-being" (1998:153). The "hell-bent" stance includes a strong element of management, as evidenced in the February 1999 federal budget in Canada which put the lion's share of increased spending on science into structures which are thought to have more controlled, determined and significantly "accountable" outcomes. The holy grail is expanded to include a presumed, but decidedly elusive, productivity benefit, and well-being beyond the physical. However, in "investing in science" with the full anticipation of returns on the dollar, "Our rulers," in the words of Sir John Kingman, "want to know in advance where science will lead, and are not pleased to be told that science does not work like that" (Kingman, 1998:8-9). Neither, of course, does the process whereby science and technology have social impacts. What Kingman (1998:8) terms the "truffle-hound principle," whereby the instincts of the best scientists are taken as the soundest guide to where new discoveries are to be found, is overthrown in favour of governments picking "winners" *a priori* or picking those areas which are thought to be most likely to produce social benefits. This is not something at which anyone has excelled, if the past be any guide.

The interest in this paper lies not in the labyrinthian relationship of the intellectual development of science and public policy, but in offering a roughly hewn overview map of the complex terrain of social impacts of technologies. In attempting this, I first look at some significant documented historical examples of social impacts, in part to underline the reality that we are not treading entirely new turf in the quest to capture social impacts. Then, I share some thoughts on how the social impacts conundrum might be usefully unpacked of some unhelpful baggage. I then outline a map, not any prematurely devised taxonomy, it must be emphasized, of how elusive social impacts might be captured as a four level process, with work proceeding on each level simultaneously and synergistically. The four levels are:

- reassembling existing data as building blocks for capturing social impacts
- developing new frameworks and ultimately new taxonomies
- working on new methods of analysis that tap into relationships among technologies and much later and deeper effects
- developing new measures of social impacts.

2. HISTORICAL EXAMPLES OF SOCIAL IMPACTS

Lest we indulge ourselves in our wit for thinking of capturing the social impacts of technologies for the first time ever, it may be wise to note some examples of capture in the past. I focus here on some select sociologically examined examples. These, by no means, are to be taken as reinforcing the "technology as destiny" doctrine. The impacts assessed have been almost universally complex and contradictory.

Our press into new technologies is paradoxically pushing our sense of human time backwards, so that we can literally dig further into our roots to understand our ancient social relations with technologies. The earliest tools thus far discovered have been dated from 2.5 million years ago. It is now recognized through the pioneering work of such notables as Richard Leakey that the earliest social impacts of technologies were on the physical construction of our own ancestral bodies. Through feedback loops, people and their tools evolved together. As tools developed for cutting meat, our "fang-like" front incisors diminished in size and utility. As agility became less required in pursuing our food on the hoof, bipedal postures developed.

Lewis Mumford (1963) saw the coordinated labour power of human beings, in building the Great Pyramids, for example, as a clear precursor to the subsequent mechanization of labour. Technologies have defined entire human epochs and cultures. Indeed, it has been recently argued, in ground-breaking research with chimpanzees in the wild, that technologies may form the bases by which animal cultures (previously argued to be a defining attribute of human beings alone) develop and differentiate. That social impacts of technologies continue in defining socio-historical eras is clear in the Fordist pact by which family structures as we know them in the 1990s developed as a function of assembly-line manufacturing, enabling families to live in nuclear groups, sharply gender-specialized, owning their own homes and cars (see McDaniel & Tepperman, 1999). It is clear from Innis' analysis of "staple industries" (fishing and the fur trade) in Canada (Innis, 1951) that communication media influence the dissemination of knowledge over time and space, and the ways in which societies self-construct and self-organize. Stones as communication media, when etched as petroglyphs for example, have enduring qualities but are not easily transported through space, thus defining a social era of stability where time is emphasized and space is not. Innis argues that time-biased communication such as stone etchings strengthen the hold of organized religion, while more portable, space-biased media such as papyrus, favour development of large-scale political institutions (Innis, 1951). Castells (1996:465, 467) extends Innis' theory into the 1990s arguing that

> Space shapes time in our society, thus reversing a historical trend...The idea of progress, at the roots of our culture and society for the last two centuries, was based on the movement of history, indeed on the predetermined sequence of history under the lead of reason and with the impulse of productive forces, escaping the constraints of spatially bounded societies and cultures. The mastery of time... transformed space in the vast movement of industrialization and urbanization accomplished by the twin historical processes of formation of capitalism and statism. *Becoming* structured *being*, time conformed space. [Emphasis in original.]

Marshall McLuhan's *The Gutenberg Galaxy* reveals how communication technologies engage humans and structure human societies in particular ways. Drawing on history, social and natural science, McLuhan traces the social transition from the "cool world" (McLuhan's now widely known term) of oral communication and handwritten books, to the "hot" era of print which he hypothesized connected people to larger, less personal social structures. Extrapolating from this, the computer and particularly the internet, usher in new kinds of communication and new social connectivities between individuals and social structures (Castells, 1996; Leiss, 1990), but as Stehr (1999a; 1999b) compellingly argues, the transition is more societally than technologically driven. That technological innovation can lead to, or be associated with, social stagnation is evident in the example of the automotive industry, as shown by Dierkes, Hofman and Marz (1998:104-107). Despite being a symbol of social prosperity (what's good for General Motors is good for America, or the comparable German example of the Volkswagen as symbol of German economic collective success), innovation in automotives tends to entrench current social problems — pollution, car-based human settlement) along with social inequalities. To capture this contradiction, Dierkes, Hofman and Marz use the term "stagnovation," a concept which makes manifest the relationship between postponing a problem and exacerbating it through technological innovation.

The "productivity paradox" (the documented disjuncture between the widespread anticipated economic payoffs of information technologies and the apparent lack of competitive return) poses similar challenges to the perception of technology as progressive in its social and economic impacts.[2] Relevant to this paper's focus on capturing social impacts, it has been argued (Castells, 1996:78) that the productivity paradox is statistical artefact, a function of "the growing inadequacy of economic statistics to capture movements of the new informational economy," a conclusion supported in part by an analysis of Canadian data (Diewert and Fox, 1999). However, Stehr (1999b) argues, in a sociological analysis of information and communication technologies (ICTs), that it is the very discourse of industrialism that mires analysts in economic assumptions about human behaviours rather than enabling understanding of current shifts in knowledge and social competencies and their profound implications for work and daily life.

Social impacts of technologies matter, not only to curious social scientists, but in illuminating our collective public choices, and in assisting corporate decisions. Understanding social impacts, broadly defined as positive, negative, or both at once, enable government and corporate decision makers better to do their jobs. Social impacts of S & T have changed our lives completely and literally, in some ways immeasurably, and play a central role in any discourse about any future we might imagine. As well, large and small-scale societal changes have transformed knowledge capacities and the ways in which science and technologies are created, developed, utilized and incorporated into processes and lives. "The world desperately needs precise data on the social impact of knowledge-intensive, digitally-oriented global production, where bits are replacing atoms (in terms of importance) in the economy" (Patterson, 1999:186), as well as in society.

3. HOW SOCIAL IMPACTS MIGHT BE UNPACKED

Capturing social impacts may not be as difficult as we presuppose. Social impacts must first be seen as worthy of study and measurement. This means that the cultural "apartheid" that has characterized science and technology, for the most part, and social sciences must be undone. This was a recurrent theme during the Science and Technology Review of the Government of Canada (June 1994-March 1996). The task in capturing social impacts of S&T is ultimately exploring an interactive relationship of science and technology *with* society. Marc Luyckx, a Brussels-based member of the European Commission's Forward Studies Unit, suggests that a transmodern view may be necessary to understand comprehensively the social impacts of science and technology. Humans now know, Luyckx suggests that "scientific discovery and technological innovation can lead not only to miracles of change but also to unprecedented dirt, damage and disease" (Das, 1999:F3). Stehr (1999b) favours a reversal in the often presumed causal direction: that postmodern social conditions open the possibilities of greater knowledge, including scientific knowledge, and to the creation of a knowledge society, rather than the more popular belief that technological changes cause (or result in) social changes. Indeed, the dialetical relationship of technology and society has long been acknowledged in the best historical analyses (see Braudel, 1967, for example). Technological determinism is seen by many as a false problem since technology is society, and societies, it is argued, cannot be understood without reference to their technologies (see Bijker *et al.*, 1987).

Unpacking social impacts may involve stripping of belief, whether heartfelt, hopeful or semi-accurate, inherent in the social impacts stances existent in the literature. Critiques of science become as valid sources for potential measures as do polyannaish technocratic futurists (Davis, Hirschl and Stack, 1997). Avoidance of both extremes in presumption and in framing the analytical questions, but not in potential source of measurement possibilities, is prudent.

The key in unpacking social impacts is the tracing back to origins of what we think may be impacts. In doing so, searching out linkages is probably more productive than *a priori* assertions of causality, which are premature without an informed theoretical perspective. In tracing, it is wise to be open to possibilities and linkages that may be hidden, improbable or contradictory (see Tenner, 1996 for some compelling examples), and trajectories that are life-like in their nuances, growth, feedback loops, and unexpected consequences. An example is provided by contraceptive technologies, which are found to "travel":

> ...while social science studies often assume that a technology performs the same function that inspires its manufacture no matter where it is applied, technologies usually undergo significant transformations... (Bledsoe, 1994:111)

Long established mechanisms by which impacts occur can be altered in their pathways by technological or social changes. An example is provided in infant mortality rates as a social mirror, long viewed as indicators reflective of socio-economic inequalities and the health impacts of social injustice. The remarkable success of neonatal intensive health care, however, argue Gortmaker and Wise (1997:155-158), has uncoupled the influence of social factors, such as poverty and inequality, on infant mortality. But the altered causal pathways, as a result of technological change and social application, are complex and multidirectional. Widespread use of new technology such as that used to control phenylketonuria (PKU), a genetic disorder, for instance, reveals that social circumstance determines the extent to which one can benefit from new technologies, with remaining disparities in incidence of PKU being purely social. New medical technologies can exacerbate social inequalities if they are not made universally available, or if there are financial barriers to access. The introduction in the United States of surfactant therapy for respiratory distress, a major cause of infant mortality due to premature birth, for example, resulted in a widening gap in infant mortality rates between African Americans and white Americans, largely because white infants benefitted more as a result of better access to the new technology (Gortmaker and Wise, 1997:157). Similarly, growing inequities have been found between socio-economic groups in access to new computer communications technologies in Canada (Dickinson and Sciadas, 1999).

Ideas may matter more to social impacts and their capture than technologies themselves. The tracing of ideas on the terrain of the social can be precarious and decidedly challenging, and yet capturing this most elusive aspect is vital to unpacking social impacts. Aristotle (as reprinted in 1970) already recognized this when he noted the five means by which truths can be affirmed or denied, with science being only one; the others are art, practical wisdom, philosophic wisdom, and intuitive reason. In late 20th century western societies, there is a tendency to privilege the truths of science and to worship at the altar of technological change as a social and economic good (Castells, 1996). At another level, the idea of technology, *and* importantly, that technology is associated with social impacts, largely thought to be good, matters greatly to capturing impacts, in both process and substance. The march of progress has paralleled for much of the past two centuries the march of industrialism along with, but not always causally related to, the ever improving quality of social life. This means, as both Stehr (1999b) and

Bell (1968) before him, have shown in different ways, that our ideas about science and technologies are captive to our ideas about modernity, goodness, and progress. Stehr (1994:251) reminds us that all societies take symbols of modernity into their traditions.

Ideas about technological change being inexorable and inherently good for society, at least in large measure, have led to a revelatory, for the purposes of impacts capture, debate about the social impacts of technology. On the one hand is the belief that the march of technological change is homogenizing and leveling of societies. On the other hand is the view, expressed well by Castells (1989:2) that technologies have "a fundamental impact on societies...but their effect *varies* according to the interaction with the economic, social, political, and cultural processes that shape the production and use of technological medium" [emphasis added]. The revelatory aspect emerges in capturing both similarities and differences in social impacts of technologies. That social impacts are socially and economically contextual is a crucial puzzle piece.

We must be clear on what we seek to know in searching out social impacts and for what purposes. It is the analytical questions that will light our way toward the capture of social impacts. It is also these questions that will ensure, partially, that we will recognize a social impact when we meet one. In capturing social impacts of technology, we are asking less about what the impact of X technology is on this or that particular social institution, as we are asking a nest of related analytical questions: Can any relationship be theorized between technological change and social change? If so, what precisely? In what direction(s) does the impact flow? Are there feedbacks? What detectable linkages exist and can be captured? What theoretical frameworks exist to explain the connections and the nature and directions of effects/relations? There are others.

4. A MAP TO CAPTURE SOCIAL IMPACTS

4.1 Reassembling Existing Data as Building Blocks for Social Impacts Assessment

An essential building block for future capture of social impacts is to assemble and assess what is before our noses. This may seem pedestrian, but it is a vital step. If there is anything learned thus far from the Y2K challenge, it is that technologies are not a universal good (although their failures can result in an economic boom) and that medium term horizons are upon us sooner than we think. It is wise to start with what we have and with that in which we have developed some confidence.

We need to move ultimately beyond sectoral approaches (studying Information Technologies, or biotechnologies, or learning technologies, etc.) in capturing the overall social impacts of technologies. In deepening sectoral analyses and social institutional impacts, however, and lengthening our tracings back to origins (to the degree this is possible in the social realm), we strengthen our understandings of social impacts of and on technologies, and of science and technology as social process. We are enabled to move beyond the sectoral by our enhanced understanding *of* the sectoral.

Linking of previously unconnected social and technology data is an important step, and one outlined as part of the strategic plan for development of an Information System for Science and Technology (Statistics Canada, 1999b) and endorsed in at least two initiatives taken by Statistics Canada in collaboration with others, the proposal to create a Social Statistics Research System, and the development of Canadian Institutes of Health. The profound challenge of connecting the social with the technological in existing data is the assessment of what influences what, and how. For an excellent discussion of these issues, see Stehr (1999b) in this volume. More analytical theory is vital for understanding these linkages and for setting out a means of data capture.

4.2 Developing New Frameworks and Ultimately New Taxonomies

It has been usefully argued that

> Progressive scholars' predictability and public credibility will remain rather feeble until we help generate serious methodologies and global data sets on variables that represent propositions derived from competing theories, subject to empirical testing...a means of substantially improving the ability to explain and predict new forms and outcomes of innovation and inequality. (Patterson, 1999:186)

The essential step in conceptualizing, perhaps most productively in competing theoretical frameworks, the complex interrelationships of technological and social change, best takes place prior to the development of taxonomies for data collection. To add emphasis to a point made earlier, we must know what we seeking before attempting to measure it. Measurement is a proxy for theoretically clarified concepts.

A site of potentially useful framework development is the well-established but less well understood, connection of technology with social stratification (Davis, Hirschl and Stack, 1997; Goyder, 1997:238;

McChesney, Wood and Foster, 1998). Although there may be aspects of framework development that would draw on existing theoretical conceptualizations such as the sociological theories of Marx (see Davis, Hirschl and Stack, 1997; and McChesney, Wood and Foster, 1998 as examples), and of Ogburn (1973), as well as of Bell (1968), other frameworks are no doubt needed. For instance, Patterson (1999:185) points out that there has been what he terms a "contextual transmogrification" whereby humans at work are now doing only what automation cannot yet do well. This, he suggests, "conjoined with impressive new science-spawned engineering design and production tools, and society has a workable prescription for class confrontation." In contrast to the predictions of Marxist theory is the empirical reality that automation-induced social crises have resulted more in worker apathy, anomie and despair than upheaval. "New globally mapped data are needed on the level of specific reasons for and ways around workers' newfound timidity..." argues Patterson (1999:186).

Old analytical data categories, such as primary, secondary and tertiary sectors of the economy, for instance, may prove to be "epistemological obstacles to the understanding of our societies" (Castells, 1996:206). A closer, harder look at the content of the existing industrial and occupational categories may be a first step in the development of new frameworks and ultimately, new taxonomies. Some analytical separation of classification schemes from societal changes related to technologies may be needed.

4.3 Working on New Methods of Analysis

New methods of analysis are undoubtedly necessitated by the interest in capturing social impacts. Given that analysis may be what we do when thinking about the problem, the more analysis, the more thought about the problem. New methods of analysis require the posing of new analytical questions and vice versa.

Adoption of longitudinal frameworks for data capture will be important. Impacts, social or economic are long-term, despite our preoccupation with how quickly life changes in the world of technologies. Statistics Canada's recently developed longitudinal surveys on children and on labour and income dynamics, for example, have placed Canada in the strongest position in the world to begin to understand the causality factors in a way that no cross-sectional study can do. Causes must be understood in order for policies, either private or public, to be developed effectively. The social impacts of science and technology ultimately can only be fully assessed by carefully designed longitudinal surveys. With the expertise existing and being developed in the Science, Innovation and Information Technology Division at Statistics Canada, and the longitudinal survey expertise at

Statistics Canada and in the social science communities in Canada, Canada is ideally situated to launch a longitudinal survey on the social impacts of science and technology.

Also vital to either future longitudinal surveys or to utilizing existing statistical data to assess the social impacts of technology, is the development of theoretical pathways that can illuminate directions in which analyses of social impacts could go. One problem in detecting social impacts empirically is the difficulty of making causal inferences from geographically aggregated data. Nonetheless, pioneering studies of socioeconomic status and life expectancies have worked around this challenge with good success. Statistical models such as LISREL and multivariate modelling have also been used to excellent ends for these purposes. And it might be that simulation techniques such as those used in Statistics Canada's LifePaths project could be adapted for following the "lives" of technologies and their social impacts.

4.4 Developing New Measures of Social Impacts

The tried and true means for improving measurement is to measure and then improve, iteratively. Questions lead to more questions, to new measures and further questions. The other option which can, and has, worked with creative people in a creative environment, is to rachet up our levels of insights into the processes involved in the production, dissemination, outcomes and social impacts of science and technology. One example of this kind of creative synergism resulted in the Statistics Canada (1999a) report, *Science and Technology Activities and Impacts: A Framework for a Statistical Information System*, which was developed as part of an intensive working group process. Most productively, both of these approaches might proceed at once, on parallel tracks. As Gilles Paquet is fond of saying, "If we don't know what we are doing, duplication and overlap may not be so bad."

We have a sufficient base to set forth bravely and audaciously asking the tough, even heretical, questions about social impacts. This is not, in any way, an evasive musing about generalities, but an analytical statement. In posing the sharp analytical questions, our way will be illuminated to social impacts capture. There is a navigational effect. What follows are a few selected examples of new concepts highly relevant to social impacts that need capture as data, and thus far, have been unmeasured or insufficiently measured. This list is illustrative of work remaining to be done in social impacts capture.

Notable is an increasing zest for measurement and experiments in measurement related to social impacts in the private sector. For example, *The Economist* notes "a new enthusiasm for better measurement" [of intangibles such as knowledge or ideas] by business ("Measuring Intangible

Assets..." 1999). Similarly, Stewart (1997:223) points out in a discussion of measuring intellectual capital, that "...enthusiastic experimentation with measurements is the best way to improve them." This growing interest of the private sector in data capture of what is essentially a social phenomenon could open the door to innovative collaborations with public statistical agencies such as Statistics Canada in developing measures.

Emerging concepts that need to be measured include, among others, social capital and human capital, stagnovation, social innovation, and that complex of terms that are said to characterize postmodern societies -- malleability, flexibility, multi-purpose resources, volatility, and heterogeneity of social structures (Stehr, 1999b). Since attention has been given to "stagnovation" as a concept above, I will focus here briefly on some compelling aspects of the other concepts as potential new measures of social impacts.

Human capital, despite being one of the leading public policy concerns in a knowledge-driven economy, argue Laroche, Merette and Ruggeri (1999:87), has been incompletely measured. This concept has become an established part of economic theory, typically referring to years of schooling only. These authors see human capital accumulation as a social activity involving the interaction of ideas and individuals that does not parallel accumulation of physical capital (1999:97). Laroche, Merette and Ruggeri (1999:98) argue strongly in favour of "a major shift in the way we look at human capital." They suggest that a useful starting point would be "to develop accurate measures of human capital stocks." Stewart (1997:232) adds force to their arguments in quoting Alan Benjamin, former director of one of Europe's leading computer services companies as saying, "The knowledge bank, not the buildings, is the reason why people invest in your company, or come to work for you." *Social capital,* a more inclusive concept, has only begun to be conceptualized (Fuyukama, 1995), and thus far has been hardly touched by data capture.[3]

Social innovation is another concept pivotal to the capture of social impacts, and one in need of accurate measurement. In short, innovations in the ways in which we live — in families, in work environments, in the build environments of cities and communities, and in our acquired capacities and attitudes with respect to new ideas and processes — matter to ways in which technologies are used, are effective, are contributory to economic and/or social progress, and to the ultimate social impacts technologies have. Yet, social innovations are not at all well captured.

Malleability, flexibility, multi-purpose resources, volatility, and heterogeneity of social structures — all define aspects, much touted, of a new social order, variously called postindustrialism, postmodernism, an information society, a network society. For the purposes of this discussion, which descriptor is used matters little. The point to emphasize is that the characterizing attributes of the new society, whatever it is called and how ever it is defined, have not been measured well as yet. Lacroix and Tremblay (1997:1), in a discussion of the long-held dream of an 'information society,' decry the "casual use of terms pregnant with meaning." The characteristics of contemporary society need operationalization, informed by existing and new theoretical frameworks, and then benchmarking so that changes can be reliably noted. Policy illumination would occur as fluctuations in flexibility in the labour force, in industries, occupations, jobs, careers and work itself could be tracked.

5. CONCLUSION

The task of this paper was to begin to develop a long-term research agenda to capture social impacts of technology in Canada. The goal was to offer a roughly hewn overview map of where the project of capturing social impacts might head. In looking at select historical examples of social impacts, several lessons are learned. We are not on new turf in this endeavour. Social impacts tend to be complex and contradictory. The relationship of technological and social changes is neither linear nor uni-directional. Neither is it universally beneficial. Impacts matter to our collective choices.

In unpacking the social impacts conundrum, several initial steps are needed: to acknowledge that social impacts are worthy of capture by statistical systems, to see that technologies develop in interaction with society and the social, to strip away myths about the social impacts of technology, to trace back to origins where possible but permitting contradictions and non-linear pathways.

In mapping out how social impacts might be captured, suggestions are offered in four categories. First, existing data could be reassembled and analyzed, particularly pulling social data together with data on technological change. Sectoral analyses could be deepened to move beyond into wider social impacts. And linked data could be developed in relation to emerging theoretical developments. Second, new theoretical/conceptual frameworks could be developed, particularly competing frameworks that could then be tested. The well established connection of technological change with social stratification could be usefully explored further in new frameworks. And old analytical categories might be usefully re-examined and changed or discarded. Third, new analytical methods could be used, particularly the development of longitudinal surveys of social impacts, drawing on recognized international leadership in the development and analysis of longitudinal surveys at Statistics Canada and in the Canadian social science community. And fourth, new measures could be sought, particularly of the key terms characteristic of the new technological society, whatever it is called. In developing these measures, the new enthusiasm of the private sector to capture social impacts might be relied upon more collaboratively.

Notes

[1] This paper has benefited from inputs from participants in the Statistics Canada-PRIME Research Workshop on "Information, Innovation and Impact," held in March 1999 in Ottawa. Particular thanks to Fred Gault for support in developing the paper and for his helpful suggestions. Thanks also to Nico Stehr and to Kwame Boadu for his help with some of the references. Given that I serve as Chair of the Statistics Canada Advisory Committee on Science and Technology, it should be noted that the views expressed are solely those of the author.

[2] See *Canadian Review of Economics*, 32(2), special issue on "Service Sector Productivity and the Productivity Paradox"; also Stehr, 1999b.

[3] I thank Nico Stehr for suggesting this important point.

PART III

MEASURING

Chapter 6

Digital Households in Canada

GEORGE SCIADAS[1]
Statistics Canada

1. INTRODUCTION

Recent technological advances, particularly in digitization, continue to lead to convergence of products, services, delivery channels and, indeed, entire industries. These developments occur in parallel with more liberalized policy and regulatory environments in most of the developed world. Leaps in telecommunications and computing especially contributed to the arrival of the *Information Highway*, which made the headlines extensively in the '90s, and heralded the entry into the *Information Society*. Today, in Canada this has given rise to the *connectedness* agenda. All these developments are predicated on the notion that "Information and its manipulation through communications networks and computers is becoming a key strategic resource that determines the competitiveness of firms and nations" (IHAC 1997, p.5). Harnessing the potential of the new technologies is widely believed to unleash a wave of creativity, which will afford businesses, governments and citizens tremendous opportunities. Innovative applications - some foreseeable, others yet unimaginable - based on state-of-the-art infrastructure, will trigger unprecedented interaction across and within sectors which "...will determine the wealth of nations and the quality of life of citizens" (OECD 1998, p.3).

As developments take place in ushering this new era, questions arise regarding the introduction of information and communication technologies

(ICTs) among all sectors, their use, as well as their economic and societal impacts. For households, pertinent among these questions have been the pace of the penetration of such technologies, the use are they being put to and the factors influencing their actual and potential use. Also of policy interest is what is happening to the *digital divide*, the gap between "haves" and "have-nots". Answers to such questions, and eventually analyses of their socio-economic impacts, will help shape future policies. Thus, connectedness metrics become necessary in order to monitor progress through quantitative research. Statistics Canada sheds light on this evolution through timely surveys and analysis.

The ICT revolution is well on its way by now. The way businesses conduct themselves vis-à-vis the consumers and each other is changing; the provision of government services on-line, education and delivery of health also represent major areas of transformed practices. While the future remains unknown, we are now at the point where some past has been accumulated. This chapter focuses on the measurement side, principally as it pertains to Canadian households. It will provide a perspective of the road that we have walked, place emphasis on where we are now and attempt to say a few things about the future. It is designed to be descriptive, rather than reminiscent or speculative, though.

Section 2, following this introduction, provides the context within which this work should be placed. It discusses statistical and analytical efforts that have taken place, explains their rationale and puts them in a macro context of national and international developments. Section 3 provides an overview of key ICT statistics among households and makes some comparisons between Canada and the US. Rates of growth and trends are identified and important findings are highlighted. In addition, it discusses the issue of the "digital divide". Section 4 examines in some detail the use of computer communications, notably the Internet, among Canadian households. It provides indicators by location of use, frequency and intensity of use, reasons for use and types of use. It also provides geographical estimates. Section 5 presents an analysis of important household characteristics and sheds light on the relation between these and penetration rates. Specifically, the relation of Internet use to income, education, age and family type are probed. Section 6 contains a brief discussion of factors with potential influence on the introduction of ICTs, as perceived by households. Finally, some concluding thoughts and direction for future work are offered in Section 7.

2. CONTEXT

The massive introduction of new ICTs started in the '80s. Businesses led the way, but by the '90s computers and modems started to find their way into households in some mass as well. However, nothing precipitated today's explosion more than the commercial arrival of the Internet in 1993. This served as the catalyst that changed perceived "normality" in many ways. Suddenly we had a whole new reality in our hands with far reaching implications, most of which we do not yet comprehend. Interest in research in this area skyrocketed. Both the private and the public sectors became active participants in this effort. One of the first things that everybody understood was the lack of a conceptual framework to put all the developments together. More than that, it became apparent that there was a lack of adequate measurements necessary to shed light into: what developments are happening; how fast are they happening, and; what are their impacts.

Many concerted efforts started to take place everywhere, in national countries and through international organizations. In the US Vice-President Gore is credited with playing a key role and much work has been carried out by the National Telecommunications and Information Administration (NTIA) among other organizations. In Canada, the Information Highway Advisory Council (I and II) was set up by the Minister of Industry. The 1995 G-7 Ministerial Conference in Brussels was instrumental in paving the way for international work. In the European Union, the Bangemann report on the Information Society was released. The Organization for Economic Co-operation and Development (OECD) conducted a series of workshops internationally on the economics of the Information Society, bringing together government policy makers, academics and businesses (Dumond and Dryden1997). Numerous other efforts are ongoing internationally, with the recent emphasis being on electronic commerce in the aftermath of the October 1998 OECD Ministerial Conference in Ottawa.

The OECD stressed the need for the development of a Global Information Infrastructure (GII) as a precursor to a move towards a Global Information Society (GIS) (OECD 1997a; Government of Canada 1996a,b). It was stated that managing "[t]he smooth and effective transition towards the information society is one of the most important tasks that should be undertaken in the last decade of the 20th Century" (OECD 1997a, p. 4, citing the Chair's concluding remarks at the G-7 Ministerial Conference on the Information Society).

In Canada, much of the brainstorming necessary to support decision making and subsequent action took place under the Information Highway Advisory Council (IHAC), phases I and II. The first IHAC report,

Connection, Community, Content: The Challenge of the Information Highway (1995), was guided by three objectives (job creation through innovation and investment; reinforcing Canadian sovereignty and cultural identity; and ensuring universal access at reasonable cost) and five principles (an interconnected and interoperable network of networks; collaborative public and private sector development; competition in facilities, products and services; privacy protection and network security; and lifelong learning) (p. vii). The federal government articulated its intention to develop a national access strategy in *Building the Information Society: Moving Canada into the 21st Century: Phase II, Conclusions and Recommendations* (1996a). The most recent IHAC report, *Preparing Canada for a Digital World* (1997), states that "[a]ccess to the Information Highway is critical to Canada's future as an information society and its success as a knowledge economy" (p. 13) and proceeds to say that an effective strategy must address at least the following three areas of public policy concern: (i) access to basic telecommunications and broadcasting services, which represent Canada's on-ramp to the Information Highway; (ii) access to new networks and services; and (iii) establishment of a formal mechanism to define access. Access to the Internet is especially singled out. The OECD refers to the explosive growth of the Internet as a harbinger of the GII-GIS (OECD 1997a, p. 5).

All the above should be considered in the context of the move towards competitive markets, the technological convergence of telecommunications, broadcasting and computer infrastructures, as well as the delivery of services and content in general (Government of Canada 1996b; Gärdin 1997; Mansell 1994).

On the policy front several initiatives started with the aim to maximize inclusion in these developments, such as SchoolNet, Community Access Points (CAPs), Smart Communities and the Health Infoway in Canada (Government of Canada 1997, 1999), telecommunications and information infrastructure assistance program (TIIAP), e-rate and others in the US (US 1999). Efforts continue on all fronts for the development of a better understanding and appropriate responses to the transformations that are taking place.

Throughout this activity special attention was given to statistical gaps, as they were rightly perceived as a major impediment to analytical work. After all, the power of information in an Information Society cannot be underestimated. This type of information is needed by all concerned: businesses, to determine the actual and potential sizes of markets - information which may impact on the introduction of new products and services requiring a critical mass, and; governments, to monitor developments and design more effective policies.

New measurements do not just happen in a vacuum. They must be done with an appreciation of their relevance, usefulness as well as potential uses. At the same time, a balance must be struck between the need for information, the use of resources and the invasion of peoples' and businesses' privacy (response burden). Therefore it was important to listen to the various potential user constituencies, to assess the feasibility of utilizing existing infrastructure for measurements or developing new instruments, and to get measurements right. Statistics Canada followed this activity closely.

Practically, efforts started in many countries. Information requirements were identified for households, businesses, governments, educational institutions and the health sector among others. Work is going on in several areas, such as business use of ICTs, availability and impact of ICT at schools, libraries and the like – these fall outside the scope of the present chapter. It was recognized that developments take place fast and that progressively questions become more interesting. In the area of households, initially the emphasis was on *access*. It was only recently that researchers were using modems as an indicator of access. Now, this has become an anachronistic indicator. We have now moved into the more interesting question of *use* and eventually we should be moving towards a better understanding of *impacts*, economic and societal.

For households, the US started their research with the "Falling through the Net" report (US 1995) and continued with "Falling through the Net II" (US 1998 - III is forthcoming in 1999). Emphasis was placed on the digital divide, with the main concern to avoid societal polarization among whites and visible minorities. Canada followed suit with both survey and analytical work. First existing data sources were tapped to produce the report "Access to the Information Highway" (Dickinson and Sciadas 1996a), followed by the "Access to the Information Highway: The Sequel" (Dickinson and Sciadas 1997). New statistical developments initially took place by adding questions related to the use of Internet in existing surveys. Then a whole new household survey was developed, specifically to deal with the issue of use. It resulted in the "Canadians Connected" report (Dickinson and Sciadas 1999) and continues with "Getting Connected or Staying Unplugged: The Growing demand for Computer Communications Services" (Dickinson and Ellison 1999). The remainder of this chapter synthesizes this work.

3. HOUSEHOLD CONNECTIVITY

3.1 An overview of selected ICTs

Telephone and cable networks, combined with more modern computer networks, are the backbone of the Information Highway. Some ICT media have been parts of our lives for a long time, while others are relatively newer. Versatile terminals, brand-new or re-incarnated, form an integral part of the revolution, reminding us of the tight vertical integration between manufacturing and the provision of services. Table 1 provides an overview of the penetration of selected ICTs among households, both in absolute numbers and rates. It also shows their recent growth history.

Changes in telephone penetration rates generally match the growth of households. Cable penetration has reached a plateau over the past few years, to also follow population growth. It continues to be the highest in the world, but other countries - especially European - use satellites more extensively. It is only recently that cable companies compete (legally) with satellite services in Canada. The last couple of years showed significant gains among households for cellular telephones, although their penetration lags behind some other countries, notably Scandinavian. Computer penetration continues to grow at high rates and the Internet, the heart of computer communications today, is increasing at healthy rates too.

In addition to increased ICT penetration at the level of the household, spending is also on the rise. This is true for the aggregate market as well as for individual households. For example, in 1996 total household spending for communications services was $12 billion – of which $7.7 billion for telephone services and $2.8 billion for cable. However, it still absorbs a relatively small chunk of family budgets with average expenditure in 1996 of: $278 for basic telephone service; $94 for enhanced telephone services (such as call waiting, call display, call forward and an increasing array of services, used by 36% of households); $435 for long distance (used by 88% of households); $457 for cellular telephony; $367 for cable; and $188 per year for the Internet (FAMEX 1998).

Table 1
Household penetration of ICTs

	1986		1996		1997		Growth Rates Average Annual Compounded 1986 - 1996	1996 - 1997
	%	000's	%	000's	%	000's	%	
Telephone	98.1	9,270	98.7	11,269	98.6	11,423	2.0	1.4
Cellular Telephone	(-)	(-)	14.1	1,611	18.6	2,152	(-)	33.6
Cable	65.4	6,177	74.0	8,451	73.7	8,539	3.2	1.0
Computer	10.3	974	31.6	3,602	36.0	4,165	14.0	15.6
Modem	(-)	(-)	15.5	1,771	21.5	2,494	(-)	40.8
Internet	(-)	(-)	7.4	844	13.0	1,500	(-)	77.7
All Households	**(-)**	**9,447**	**(-)**	**11,412**	**(-)**	**11,580**	**1.9**	**1.5**

Source: HFE (1997).

Note: (-) indicates that comparable data are not available

4. COMPARISONS WITH THE US

The US recently completed a second round of their "Falling Through the Net" study (US 1998a), which is used as the basis for the present comparison, in conjunction with the Household Internet Use and the Household Facilities and Equipment surveys in Canada, as well as some other sources (identified in Table 2).

Table 2
Canada - US comparisons, 1997

	Telephone	Cellular	Cable	Computer	On-Line	E-mail
	%	(per 000 inhabitants)	%	%	%	%
Canada	98.6	138	73.7	36.4	16.0	13.3
United States	93.8	207	67.3	36.6	18.6	16.9

Sources: HFE (1997), Household Internet Use Survey (1997), US (1998) and author's compilations
Canadian data for telephone and cable come from the May Household Facilities and Equipment Survey. Data for computers, computer communications and e-mail come from the October Household Internet Use Survey. US data for telephones, computers, on-line access and e-mail come from the November "Falling Through the Net" Survey, while data for cable come from the American Cable Association. Canadian and US cellular telephony data come from Chodorowicz and Sciadas (1998).

Comparisons are indicative only, as they are affected by several factors, including sampling and timing differences. This time, though, data for computers, computer communications and e-mail are more comparable as the timing of the US and the Canadian surveys was almost the same (Fall 1997). This is important in an area where a few months can make a noticeable difference.

Telephone and cable penetration rates in Canada are higher than in the US, but the opposite is true for cellular phones – although no direct comparison is possible at the household level. Computer penetration is comparable, whereas more American households had on-line access and used e-mail compared to Canadian households. These overall averages, however, conceal significant variations by income, education, geographical location and other characteristics, which only a detailed analysis, beyond the scope of the present work, could uncover.

4.1 The digital divide

While penetration rates are increasing across the board, of particular policy interest is the degree to which this is happening among different population groups, indicative of the evolution of the gap between those who are connected and those who are not.

Comparative analysis of this kind involves several nuances. For one, comparisons between "haves" and "have-nots" are really comparisons between the most-connected and the least-connected groups and not the not-connected. Table 3 presents an analysis of the evolution of the gap, based on the use of the Internet between 1996 and 1997. It compares the, analytically instructive, extreme penetration population segments, the highest vs. the lowest income quartiles, the highest vs. the lowest education level, the most connected vs. the least connected age group and Census Metropolitan Area (CMA) vs. non-CMA locations.

Table 3
Comparative analysis of Internet growth

	1996	1997	Change in gaps*	Rates of growth
Income Quartile				
Bottom	2.7	5.3	(+) 6.8	100.2
Top	15.2	24.6		64.4
Education				
< High School	1.0	2.7	(+) 7.9	159.1
University +	18.6	28.2		58.0
Area				
Non-CMA	4.5	9.2	(+) 1.4	110.6
CMA	9.7	15.8		66.2
Age				
65+	2.1	3.7	(+) 5.5	80.4
35 - 54	9.8	16.9		77.8

* Difference in penetration rates between high and low penetration groups in 1997 minus their difference in 1996

1. Penetration rates have increased across income quartiles, levels of education, age groups and geographical locations. In that sense, the gap is closing as more people are connected and fewer are not. That is, some of last year's "have-nots" have turned into this year's "haves". A more meaningful question to address in order to get a better understanding of what is happening to the gap is what population sub-groups did the new

"haves" mostly come from. Here we must consider both absolute and proportionate changes.

2. Simply comparing the difference in penetration rates over time reveals that penetration is increasing by more percentage points for the high rather than the low-penetration population segments. This is shown in the "change in gaps" column of Table 3. For instance, the gap in penetration rates between the top and the bottom income quartiles was 6.8 percentage points greater in 1997 than in 1996. In that sense, the gap is "widening". Similar analysis in the US found that this gap is widening, not only for lower compared to upper income groups, but also for the less educated, the youth and minorities (US 1998a). These findings show that, in absolute terms, more of the newly connected households come from the have segments. However, such findings may mask important underlying growth trends.
3. Penetration rates and their differentials are driven by connectedness rates of growth among population sub-groups. What happens to the gap, depends crucially on the magnitude of the initial mass of connected households, the associated gap in penetration rates by population sub-groups, as well as the respective rates of growth. Higher rates of growth among low-penetration groups may still result in a widening gap. For instance, starting from a much higher number of connected households, high- penetration groups can gain one percentage point in penetration with only a small percentage increase in user households, whereas low-penetration groups need a much higher rate of growth. To the extent that this is the case, though, it will be indicative of future trends. The underlying growth trends are unmistakably much higher among the low-penetration population segments (last column in Table 3).

Alternatively, this can be demonstrated by the likelihood of individual households being connected. For instance, in 1996 households in the top income quartile were 5.6 times more likely to use the Internet than households in the bottom income quartile, while in 1997 they were 4.5 times more likely. These trends are true for all the other sub-groups but they are proportionately more pronounced among education groups, followed by location and income and much less by age group.

More detailed analysis reveals that not only this pattern is maintained, but it is stronger. For instance, stratifying sub-groups by income and education combined, we found that among the least educated who also were in the lowest income quartile, the rate of growth was 307%, compared to 53% for those at the highest income quartile and the highest level of education.

Preliminary analysis of 1998 data generally confirms these findings. An exception has been noted in the income variable, where households at the top income quartile experienced a higher rate of growth than those in the lowest income quartile between 1997 and 1998, in addition to a widening gap in penetration rates (although that gap was reduced from the 1996-1997 one). Continuous monitoring of this situation is warranted.

5. USE OF COMPUTER COMMUNICATIONS

Computer communications refer predominantly to the use of the Internet, with the exception of limited electronic banking, which may be taking place over closed networks. This section examines in some detail use among households utilizing data from the October 1997 and 1998 surveys. It takes advantage of the opportunity afforded by the recent data to offer some trend analysis as well.

5.1 General indicators

The proportion of households containing at least one member that has ever used computer communications from any location continues to increase and in 1998 it reached 45.7% of all households, up from 38.1% in 1997. However, although a sizeable market already, the Internet is nowhere close to being the ubiquitous medium popular hype would have some believe. Regular use from any location reached 4.3 million households in 1998, up by 843,000 (24.6%) from 1997. The analysis that follows relies exclusively on regular users.

Although many households use from multiple locations, the most prominent location of use is the workplace, followed by the home. Home penetration is substantial given the relatively short life of such technologies. At the same time, the data show a closing gap between home and work use from 1997 and 1998. School is the third most important location of use, while only a small proportion uses from public libraries and other locations, such as cybercafés.

The regular use rate increased for all major points of use (home, work, school and public library) (Chart 1). The biggest increase between 1997 and 1998 is for use from home. It rose by 41%, with a slightly bigger percentage increase in the number of home user households (43.9%). The work use rate rose by 16.6%, and the number of work use households increased by 18.9%.

The use rate for use from school showed the second-largest percentage increase (28.9%), with the number of households increasing by almost one-third (31.4%). The use rate from a public library remains low with little increase on the 1997 rate. Even so, when combined with the rise in the total number of households, the number using from a public library increased by

20.6%. The rate of use from other locations showed a marginal decline, perhaps attributable to the greater incidence of use from home.

Chart 1: Use of computer communications
(all households, various locations)

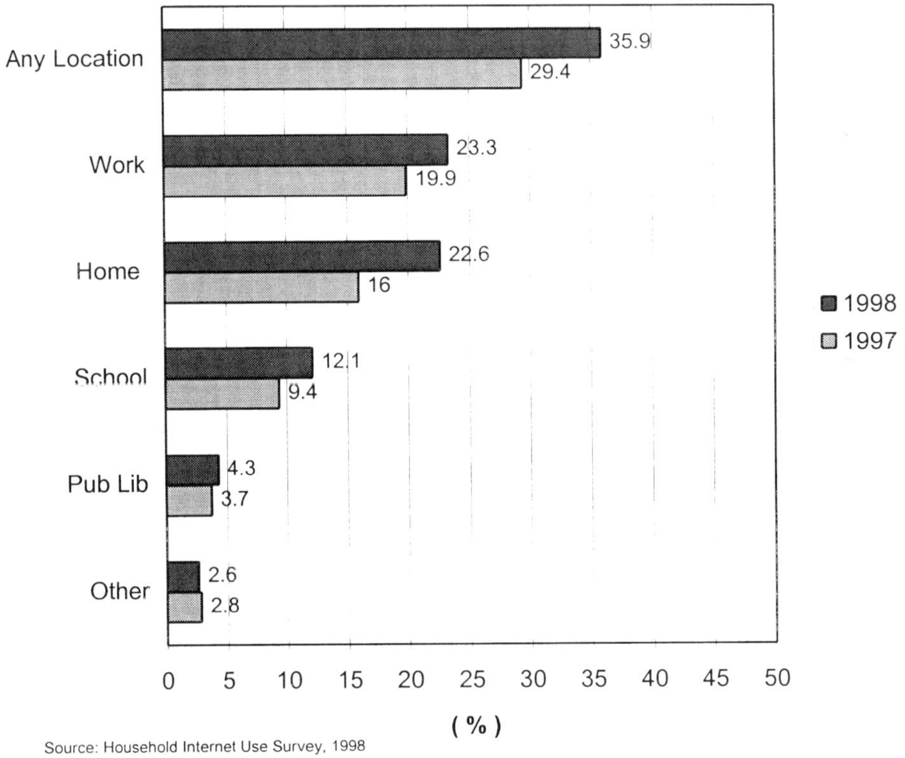

Source: Household Internet Use Survey, 1998

Beyond the incidence of use, however, of interest is the degree of use, the reasons for use and the type of services used. In terms of the frequency of use, there has been little change in the proportions of regular home user households averaging daily, weekly and less than weekly use from home. Among users, the proportion averaging daily use (62.2%) continues to dominate, while only a very small share (4.6%) average less than weekly use. Adjusting for the increase in the home use rate, 14% of Canadian households were averaging daily use from home in 1998, compared with 9.8% in 1997.

The intensity of use by regular home user households also increased a little, with a slight increase in the share of heavy users (20 or more hours per month) and a slight decrease in the share of light users (less than 10 hours). They continue to dominate medium users (Chart 2). Factoring in the increase in the overall home use rate, one in ten Canadian households (9.9%) used heavily from home in 1998 (up from 6.7% in 1997). 8.2% were light users and 4.3% were medium home users.

Chart 2: Types of users by hours of use per month, 1998

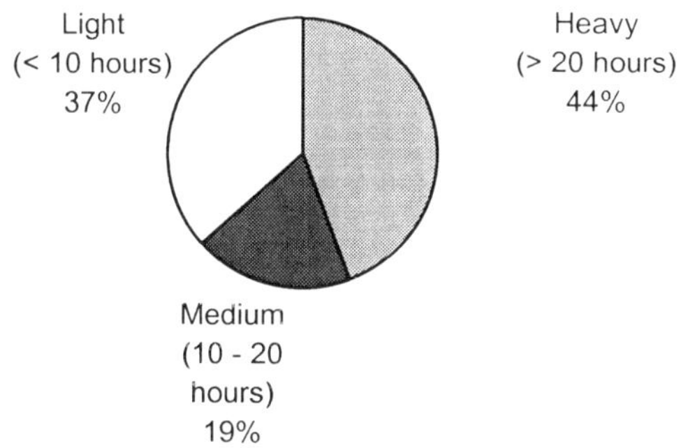

Source: Household Internet Use Survey, 1998

Among the list of choices in the surveys, the dominant types of use are e-mail, general browsing and searching the World Wide Web for specific information. (Table 4) The second-highest range includes using the Internet for medical and health information, government information, playing games, and using for formal education or training. These are followed by participation in chat groups, electronic banking and purchasing of goods and services.

Although electronic banking has a low penetration rate among home user households, it did increase by 19.6% since 1997. Furthermore, the number of households doing electronic banking from home increased remarkably, by 79%. Similarly, the number using the home computer to purchase goods and services increased by 71.4%.

Table 4
Purpose of home use, 1998

	Penetration rates			
	% of all households using	% of regular home user households using	Percentage of households spending < 10% of their Internet time on this activity	Percentage of households spending < 10% of their Internet time on this activity
E-mail	19.3	85.6	36.8	80.5
Electronic Banking	5.5	24.4	67.0	90.0
Purchasing	2.5	10.9	91.6	99.3
Search for medical info	9.6	42.5	74.8	97.2
Education / training	6.8	30.0	32.3	79.9
Look for government info	8.2	36.4	75.9	97.9
Look for other information	15.3	67.9	28.5	80.1
General browsing	17.6	78.1	29.9	80.0
Playing games	7.8	34.4	48.4	90.1
Chat Groups	5.7	25.4	52.4	88.4
Other Internet services	2.6	11.6	71.7	96.5

Among the 6.8% of Canadian households that use from home for formal education and training, 83.1% do so for project assignments and/or academic-related research. Between one-quarter and one-fifth use for distance education (21.6%), which includes self-directed learning and correspondence courses, and to communicate with teachers and/or peers (24.5%). 15.5% use for other formal education and training purposes.

Canadian households also appear to have become more diversified in their use of computer communications from home. There is no one service on which more than 80% of the users of that service spend more than half their total home use time. Only one-tenth or less of users of each service spend more than half their time on electronic banking, purchasing goods and services, searching for medical/health information and for government information, playing games, and other services.

Furthermore, there is no single service on which significantly more than 60% of users of the service spend even one-quarter of their total home use time. And there are only four services (email, formal education and training, searching for specific information and general browsing) on which significantly more than half the users of each service spend even one-tenth of their total time.

Almost all regular home user households use the Internet for personal non-business reasons (95.2%), nearly one-third (31.7%) use for employer-related purposes, and nearly one-quarter (22.9%) use for self-employment purposes. These proportions changed only slightly since 1997, and the large increase in the home use penetration had the biggest impact on the number of households using for each purpose. The number increased by about one-half for self-employment use (51.2%) and personal non-business use (48.8%), and by more than one-third for employer-related use (35.2%).

Personal non-business use continues to take up the largest share by far of total home use time. Among those using for this purpose, almost three-quarters (73.4%) spend more than half their time on personal non-business use, more than half (54.6%) spend at least 90% of their time, and 38.3% spend all their home use time for this purpose.

Among households that use for work-related reasons, there has been a noticeable reduction in the share of total home time used for employment purposes. Among those using for self-employment, the proportion spending less than half their total time for this purpose increased (from 53.5% to 62.2%), while the proportions spending all their time (6.7%) and at least 90% of their time (13.6%) fell a little.

Among households using for employer-related reasons, the proportion using less than half their total home use time for this purpose increased (from 62.5% to 73%), and the proportions spending all their time (4.1%) and at least 90% of their time (10.1%) declined marginally.

Table 5
Use of computer communications,
by location and province, 1998

	Location of regular use					
	Home	Work	School	Library	Other	Any
Canada	**22.57**	**23.25**	**12.05**	**4.32**	**2.62**	**35.86**
Newfoundland	15.39	17.20	14.54	4.67	3.66	28.86
PEI	17.11	20.97	14.51	4.49	(2.61)	35.40
Nova Scotia	23.71	22.87	17.44	5.58	3.36	37.92
New Brunswick	18.16	19.81	12.25	2.33	(2.10)	31.02
Quebec	15.60	16.38	7.60	2.83	1.70	26.23
Ontario	25.50	25.54	12.59	4.74	2.37	39.04
Manitoba	19.94	22.27	12.53	2.41	2.75	33.49
Saskatchewan	18.26	21.51	13.93	3.66	2.55	33.75
Alberta	27.71	31.71	17.58	4.82	3.67	45.06
British Columbia	27.91	26.72	13.09	6.45	4.17	41.95

Note: ( ) indicates low reliability estimate due to sample size.

Of course, these changes do not mean that work-related use is becoming less prevalent. First, the large percentage increases in the numbers of households using for self-employment and employer-related purposes show that the incidence of work-related use from home has increased significantly. Second, even within any one household, a fall in the share of time need not imply a fall in the absolute amount of time. The work-related user may be devoting an increase total time to personal non-business use, and other members of the households may have started to use the home computer for personal activities.

5.2 The geographical dimension

Significant regional differences exist in the pattern of use. Alberta, British Columbia and Ontario top the list, followed closely by Nova Scotia (Table 5). Quebec has the lowest use rate; it does represent, though, the second largest market with almost 810,000 households using from any location, due to its population. The size of the home-use market falls short of half a million households. On either count, the absolute size of the second largest market is less than half that of Ontario.

Between 1997 and 1998, household use rates increased in every province. Penetration rates from any location of use remain above the national rate in Alberta, British Columbia, Ontario and Nova Scotia, with Quebec remaining at the lowest rate. The ranking of other provinces altered from 1997 to 1998, as the penetration rates for P.E.I. and Saskatchewan moved closer to the national rate. P.E.I. had the biggest percentage increase (62.7%) in its home use penetration rate, but it contains a tiny proportion of all Canadian households.

The percentage increase in the absolute number of households using from any location was biggest in P.E.I. (37.2%), Alberta (33.9%) and Quebec (32.5%). Ontario continues to have the largest share of Canadian households regularly using from any location of use, with Quebec coming second. However, between 1997 and 1998, Ontario's share of all regular use households fell (from 41.6% to 40.2%) while Quebec's share increased (from 17.8% to 18.9%).

Quebec also has the second-largest percentage increase (53.3%) in the home use rate. Because Quebec also has a large share of all Canadian households (25.9%), the percentage increase in the number of regular home use households in Quebec (55.7%) was substantially higher than in Ontario (35.5%). Nevertheless, Ontario continues to have the highest share of all regular home use households (41.7%), and Quebec has the second-highest share (17.9%).

In all provinces the percentage point increase in the home use rate exceeded the increase in the work use rate. In the three most populous provinces (Ontario, Quebec and British Columbia) the difference between the home and work use rates is one percentage point or less.

Differences in use rates are even more pronounced among Census Metropolitan Areas (CMAs). The fifteen largest CMAs contain more than half (56.3%) of all Canadian households. Taken as a group, households in these CMAs have higher penetration rates from all major points of use than do households outside these CMAs. However, not only did the penetration rates increase from both groups between 1997 and 1998, but also the percentage point increases were remarkably similar (Table 6).

Table 6

Use of computer communications, by location and CMA, 1998

	Home	**Work**	**School**	**Any**
Canada	**22.6**	**23.2**	**12.0**	**35.9**
Halifax	33.8	33.7	21.5	50.2
Quebec	21.1	18.6	(8.3)	28.6
Montreal	18.2	20.6	8.0	31.6
Ottawa	36.2	42.3	15.8	55.3
Toronto	27.7	29.0	12.5	42.0
Kitchener-Waterloo	25.9	26.3	17.1	42.4
Hamilton	26.8	28.1	14.1	41.2
St Catherines - Niagara	19.2	(14.5)	(9.4)	29.3
London	29.0	24.7	(15.3)	40.4
Windsor	(15.7)	(14.3)	(11.0)	26.8
Winnipeg	23.6	26.2	13.5	37.8
Calgary	34.8	38.6	20.2	52.8
Edmonton	27.0	30.7	13.9	43.9
Vancouver	31.6	30.5	13.6	45.7
Victoria	34.5	27.9	(14.9)	48.5
Largest 15 CMAs	26.2	27.3	12.5	40.3
Others	17.9	18.1	11.5	30.1

Note: ( ) indicates low reliability estimate due to sample size.

With one exception, penetration rates from the three major points of use increased in all fifteen CMAs. The exception is Ottawa, where a small increase in the home penetration rate and a small reduction in the work penetration rate left the overall penetration rate for use from any location practically unchanged. This is the first ever sign of a plateau in the monitoring of these statistics. Nevertheless, Ottawa remains the most connected CMA, whether from home, work or any point of use. Despite the slight reduction, its high work penetration rate remains noticeably above that of second-highest Calgary, and significantly above third-highest Halifax. In each of these three CMAs, more than half the households now regularly use computer communications from some location.

The increase in home use rates actually exceeded ten percentage points in five CMAs - Halifax, Hamilton, London, Calgary and Victoria. Halifax also saw a six-percentage point increase in its school use penetration rate, which is now the highest in the country. The percentage increase in the number of households using from home was highest for Halifax (76.8%), Quebec City (70.8%) and Hamilton (69.2%). Not surprisingly, the largest shares of all households regularly using from any point are in Toronto (16.6%) and Montreal (11%).

There may be a combination of reasons explaining these large increases. For instance in Halifax there was a substantial increase between 1997 and 1998 in businesses with no employees, from 7,737 to 13,332 according to the provincial business directory.

The geographical pattern of hours of home use is more or less similar to the national pattern, with significantly higher proportions of heavy and light users than medium users. The most notable exception is Quebec with an elevated proportion of light users, whereas Ontario has more heavy users. There are only minor geographical differences in the proportions of regular home user households that use for self-employment and for personal non-business reasons. There are no significant differences in the type of use either.

6. HOUSEHOLD CHARACTERISTICS

The use of the Internet is related to a variety of factors, including costs, ease of access and use and the availability of useful applications. (The latter both affect and are affected by use). Other factors reflect the demographic characteristics of households. The following such characteristics are examined: income; education; age, and; family structure. The examined relationships are correlations and do not necessarily imply causation, especially since the demographic groupings are not mutually exclusive.

6.1 Income

The very strong relationship between household penetration rates and income has been amply documented (Dickinson and Sciadas 1996, 1997). This relationship exists regardless of point of use. For use from any point, the penetration rate in the top quartile is five times the rate in the bottom quartile. The difference is more than six times for use from home and more than twelve times for use from work, but falls to between three and four times for school use (Table 7).

Table 7
Use of computer communications, by location and income, 1998

	Home	Work	School	Library	Other	Any Location
All Households	**22.6**	**23.3**	**12.1**	**4.3**	**2.6**	**35.9**
Bottom Quartile	7.1	4.1	6.1	2.7	2.0	13.2
Second Quartile	13.6	12.1	7.9	3.2	2.3	23.6
Third Quartile	24.4	26.3	13.1	4.9	2.5	41.5
Top Quartile	45.1	50.4	21.0	6.6	3.7	65.1

Because of the low penetration rates, the bottom quartile of the income distribution accounts for less than one-tenth (9.2%) of households using from any point of use, 7.9% of households using from home, and 4.4% of households using from work. Even for school use, the bottom quartile makes up only 12.7% of school use households. This is in stark contrast to the top quartile of the income distribution, which makes up at least one-half of home use households (50%) and work use households (54.3%). It also accounts for more than 80% of households using from school (43.7%) and from any point of use (45.5%). Only the third income quartile shows a degree of consistency between its share of all households and its share of user households, accounting for between 25%-30% of households using from each and every point of use.

Not only do penetration rates rise with income, but the percentage point increase since 1997 in penetration rates from the major points of use also rise with income. The penetration rate for work use in the bottom quartile actually fell between 1997 and 1998, but the drop (of less than one percentage point) is sufficiently small that statistical considerations caution against attaching much significance to it.

While the link between affordability and income doubtless explains some of the relationship between income and penetration rates, the extent of any causal relationship between affordability and penetration rates is not known. Two examples will illustrate this point. First, affordability is not an issue for work use, since costs are carried by the employer and not the household. Second, at today's prices, affordability certainly can not explain why three-quarters of households in the third quartile and more than half the households in the top quartile do not use from home.

6.2 Education

Use rates increase substantially with the level of education. Between 1997 and 1998 these rates increased for all education groups, but the same relationships remain. The use rate from any location for households where the head has a university degree is almost twice the rate for those with a high school or college qualification, which in turn is three times the rate for those that did not complete high school (Table 8).

Table 8
Use of computer communications, by location and education, 1998

	Home	Work	School	Library	Other	Any Location
All Households	**22.6**	**23.3**	**12.1**	**4.3**	**2.6**	**35.9**
< High School	6.6	5.6	5.3	1.8	1.2	12.6
High School / College	23.1	23.0	13.0	4.3	2.9	37.4
University Degree	46.7	52.5	19.9	8.5	3.8	68.1

A similar pattern, with slightly different multiples, is seen for home and work use rates. Indeed, for all specific locations of use, the use rates for those with a high school or college qualification are very close to the use rates for all households. Thus the low rate of the least-educated group is basically offset by the high rate of the most-educated group.

The percentage point differences for the school use rate are much smaller than for other points of use, but the school rate for the lowest education group is still very small. One possibility could again be related to age, if the proportion not completing high school is significantly higher among older families whose children have left home and/or are above school age. This is an avenue for further research.

6.3 Age

Use rates from major locations increased for all age groups, but there remains a substantial difference between younger and older groups. The use rate from any location remains highest for the 35 to 54 age group, although the under 35 group is a very close second. The use rate drops substantially for the 55 to 64 group and then there is another substantial drop to the 65 and over age group. A similar age pattern is seen, albeit at lower penetration rates, for use from home, work and school (Table 9).

Table 9
Use of computer communications, by location and age, 1998

	Home	Work	School	Library	Other	Any Location
All Households	**22.6**	**23.3**	**12.1**	**4.3**	**2.6**	**35.9**
Less Than Age 35	26.1	30.2	11.3	4.8	1.3	45.3
Age 35 to 54	30.1	31.4	19.2	6.1	3.1	46.9
Age 55 to 64	18.2	16.3	6.8	3.1	1.6	27.5
65 and Over	5.3	2.6	(0.8)	(0.7)	(0.5)	7.2

Note: ( ) indicates low reliability estimate due to sample size.

Although the home use rate is a little lower for the under 35 group than for the 35 to 54 group, the younger group has a higher penetration rate in all but the top income quartile. It is the effect of the top quartile, therefore, that makes the overall penetration rate higher for the older age group. In the top quartile, not only does the older group have the higher penetration rate (48.4% vs. 44.1%), but also they make up 62% of all households in this quartile. The younger group, on the other hand, makes up only 19.3% of all households in the top quartile. Also, the share of the age group that is in the top quartile is higher for the 35 to 54 group (35.1%) than for the under 35 group (22.1%).

Because the 35 to 54 age group has the highest penetration rates and contains the largest share of all households (44.3%), it makes up 57.9% of households using from any point and 59% from home. In contrast, households headed by a senior are one-fifth of all households (20.3%) but only 4.1% of those using from any point and 4.8% from home.

The relationship between age and income distribution has a noticeable effect on the penetration rate in the bottom income quartile. Households with heads aged 55 and over are one-third (33.8%) of all households, but one-half (50.7%) of households in the bottom income quartile. When they are dropped from the calculation, the penetration rate in the bottom quartile rises from 7.1% to 12.4% for home use, and from 13.2% to 23.1% for use from any point.

6.4 Family type

Rates of use from home, school and any location increased significantly more for single-family households with unmarried children under 18 than for other family types. Nearly one-half (47.6%) of single-family households with such children use from any point, replacing multi-family households as the group with the highest penetration rates from any location of use, as well as from home and from work (Table 10).

Table 10
Use of computer communications, by location and family type, 1998

	Home	Work	School	Library	Other	Any Location
All Households	**22.6**	**23.3**	**12.1**	**4.3**	**2.6**	**35.9**
SF, Children	31.0	30.0	22.8	6.7	3.5	47.6
SF, No Children	22.5	22.4	8.4	3.6	1.9	34.2
One Person	10.5	14.3	2.2	1.8	1.9	20.4
Multi-Family	26.2	28.3	14.5	6.2	5.6	45.5

Note: ( ) indicates low reliability estimate due to sample size.
SF = Single Family with unmarried children under age 18

Multi-family households continue to have the second-highest penetration rates, but they account for only 4.4% of all households. In fact, multi-family households fell a little both as a proportion of all households and in absolute numbers. It is quite possible that this merely reflects the older children of some baby-boom families leaving home, thereby reducing the number of multi-family households and increasing the number of other household types.

In 1998, more than one-fifth (22.8%) of single-family households with unmarried children under 18 used from school. This is almost double the overall school use rate (12.1%), and is a much more realistic indicator. Indeed, an avenue for future research could provide an even more meaningful school use rate if the population against which it is calculated were further restricted to include only households with school-age children.

7. ATTITUDES TOWARDS FUTURE USE

Undoubtedly, many more households will use the Internet in the future due to easier access, falling costs and more useful applications. Questions regarding what might influence their use were asked in the surveys. They can be categorized into four sets: households that regularly use from home were asked what would increase their use from its current level; households that do not use from home but do have a home computer were asked why they do not use it for accessing the Internet; households without a home computer were asked what would induce them to start using, and; finally, all home non-user households were asked whether and why they would use from a public library or other public place if easily and cheaply available.

When reviewing the data from these questions, it should be remembered that they result from the subjective opinions of one member of a household that may contain several members. Also, because of the subjective nature of the questions, the precise interpretation of some questions may differ across respondents. For example, how "much lower" would costs have to be to induce home use, and what type of "more and better services" would need to be offered. The latter issue also raises the additional problem that many non-user respondents may well be unaware of the services that are or would become available. How many of those saying nothing would induce them to use would change their minds when they learned of the services actually available? Nevertheless, peoples' attitudes are summarized below.

Current users were asked how much their use would increase if costs were much lower, if more and better services were available, and if access were easier (though, for example, via the television screen). The majority of regular home user households reported that there would be little or no change in their use, whether the stimulus was much lower costs (74.8%), more and better services (64.1%), or easier access (65.6%). The share reporting that there would be a substantial increase in home use was largest for easier access (12.6%), followed closely by lower costs (11.1%) and improved services (10.7%). The share reporting a noticeable increase was largest for improved services (24%), followed by easier access (20.7%) and lower costs (13.1%).

Of the more than three-quarters (77.4%) of households than do not regularly use computer communications from home, almost one-quarter (23.3%) do have a home computer. When asked to give the main reasons why they do not use their home computer for accessing the Internet, more than one-third (34.6%) said it was too costly, 13% said they use from work, while lack of useful services was cited by less than one-tenth of these households. One in twenty cited use from alternative locations a reason. As to what would induce them to start using, their attitudes are summarized in Chart 3.

Chart 3:
Inducements to use (non-regular users with home computers), 1998

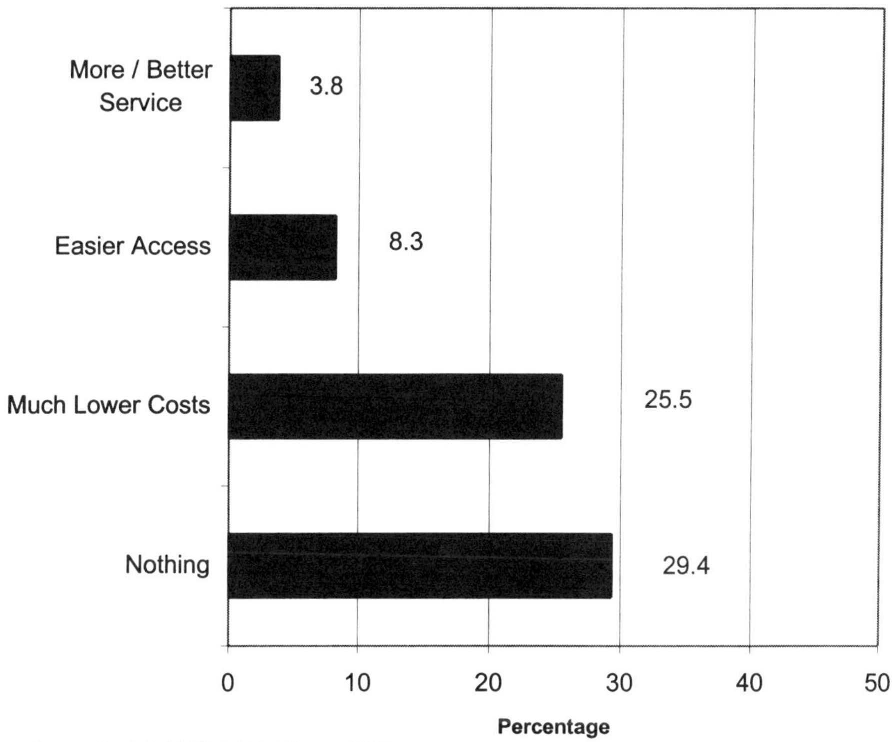

Source: Household Internet Use Survey, 1998

Non-user households without a computer were asked what would induce them to start using from home, the largest proportion (38.3%) said nothing would do so. The major inducements to use would be much lower costs (33.3%) and other unspecified factors (29.4%). Only a very small proportion said they would respond to easy access via the television set or to more and better services being available.

Of the 9.2 million households that do not regularly use computer communications from home, one-quarter (25.9%) would use from a public library or other public place if the service was easily and cheaply available. As to the issue of whether cost or ease of use would be the major determinant, two-thirds (65.1%) of the households that would use feel the two are of about the same importance. Another one-fifth (19.2%) put greater emphasis on ease of use, while 15.1% put greater emphasis on the cost factor.

8. CONCLUSIONS AND FUTURE WORK

Canada is well positioned, with international standards, to enter the Information Society, as has been repeatedly documented (see, for instance, OECD 1997b; Government of the U.K. 1996). It possesses world class telecommunications and cable infrastructures, which continue to be updated and expanded. These are complemented with new high-speed, high-capacity networks, which are being built to provide the bandwidth on which to base the offering of new applications and services.

At the same time, general connectivity increases, reaching deeper into citizenry. We are now beginning to understand the use of ICTs and the factors influencing it. The deployment of such technologies has profound implications for our family and social lives. The way we work, learn and play is in a transition phase. The way we interact among ourselves is one example. Time use and substitutions among activities is another. In Canada many users use the Internet daily and more than 40% of Internet users spend more than 20 hours a month on this new medium. At the same time Canadians spend on average 20 hours a week viewing television and 21 listening to the radio. Substitution of time use among these and other media are to be expected, as is parallel and simultaneous engagement. Complementarities may be found as convergence takes full effect. Much can be learned by studying the characteristics of those households that are on their way to taking advantage of the new opportunities afforded by the domestication of information and communication technologies and services. In fact, it has been amply emphasized that the consumption of such technologies and services seriously influences their design and production

(Silverstone and Haddon 1996). There are competing visions of the future effects of the introduction of such technologies (see, for instance, Negroponte 1995, Stoll 1995). Much more research is needed in order to begin to understand how inputs to economic and social activities are linked to various outcomes and impacts (Gault 1997).

Another issue to be dealt with is that of Internet dropouts. Usage statistics, including the ones in this chapter, have not addressed this issue. Analysis of the subject (Katz and Aspden 1998) reveals that this is not a trivial phenomenon. We need to look at and profile this group of former users, in addition to the current users and non-users. This group may partly account for the difference between those who have ever used computer communications and regular users. In reality, because of the dropouts, the number of new users is greater than the change in current users between time periods. Attempts should be made to incorporate this into future survey and analytical work.

In addition, we now have the whole new and emerging area of electronic commerce, in its business-to-household dimension. Households interact with businesses to obtain information and purchase goods and services – both ordering and making payments. Issues surrounding this area warrant special attention due to their intricacies and implications.

Of course, work on non-household related areas, such as business-to-business electronic commerce, and development of a better understanding of governments on-line, the delivery of education and health services as well as many others is essential.

Notes

[1] This chapter draws heavily on work done over the past four years. Many thanks are due to Professor Paul Dickinson of McGill University who has done much of the detailed analytical work. Most of the statistical analysis here can be found in our joint papers. Thanks are also due to Fred Gault who has been supportive throughout this work and many colleagues in Industry Canada and Statistics Canada who have collaborated in these developments in many ways over the last few years.

Chapter 7

The Use of the Internet and Electronic Commerce in the Canadian Banking and Insurance Industry

DAOOD HAMDANI
Statistics Canada

1. INTRODUCTION

The Internet is the defining technology of the information society. By reducing the cost and increasing the speed at which the information can be collected, stored and analyzed, it is creating a new business environment, changing the nature of the firm and extending the boundaries of the marketplace. The Internet is a highly efficient sales and delivery channel and offers vast opportunities to introduce new products (goods or services) and better substitutes for existing products. And it is becoming more efficient, more accessible and easier to use, making it an increasingly more attractive place to transact business, connect with suppliers, customers and partners, mine information, innovate and create wealth. The non-proprietary nature of the system and its global architecture invite and attract continued improvement and growing applications to accommodate new ideas and products.

Its reach is pervasive. By transcending national boundaries, it creates competition where only local players previously mattered or monopolies dominated. Hardly any industry remains untouched. The service industries whose products are in digital form and consumed in the digital domain such as music, software and financial services are the most affected.

Delays in adopting new technologies could be rationalized in the past on the grounds that it allowed firms to use their existing physical assets, if large amounts of money had been invested in them, to their full economic life before writing them off, with the expectation that in the meantime a new technology would emerge that would be better than what the competitors had. This worked when the life cycles of products and technology were relatively long and technological change and exchange of products respected the national boundaries. Being late in today's business world in adopting the latest electronic technology is to run the risk of never being able to catch up, or even of being cannibalized by the competitors. For the digital technology's primary effect is not the change itself but the rate of change. Just as the firms begin to understand the marketplace and manage the uncertainty a new competitor, a new product, a new production process or delivery channel emerges. The change is taking place faster than an average firm's ability to respond using traditional methods.

Once relegated to the position of an enabler for the implementation of business strategy, technology's role has become more important with the advent of the Internet as business change originates with digital technology (Downes and Mui, 1998).

The traditional business and economic indicators are not adequate to monitor or measure the progress of this new business phenomenon. Development of new concepts and measurable definitions of the changing business environment is a new endeavour. The early efforts at measurement in Canada have concentrated on the household use of computer communications including the Internet and provide measures of the type of use, frequency and intensity.

The present paper addresses the business side. It reviews the state of the preparedness of the Canadian business to take advantage of the opportunities the new technology offers. The focus is on the financial services industry because it is more amenable to transactions on the Internet than many other industries, this industry is known as well as under competitive pressure to introduce the leading edge technologies, and many of its customers are used to doing business electronically.

The study presents for the first time findings from a survey. It provides estimates of the proliferation of the Internet and selected uses including electronic commerce to shed light on the digital strategies of the Canadian financial intermediaries. Estimates are given of firms that sell their products on the Internet as well as of firms that do not sell on-line but conduct business-related activities that may eventually lead to a transaction. As well as providing estimates to indicate how far along the Canadian financial services industry is, an analysis is presented of how the adoption of information technologies affects firms' ability to innovate which lies at the heart of the competitive advantage of firms and nations and drives economic growth.

2. CONNECTEDNESS: THE BUSINESS SIDE

The mix in which businesses combine their human and physical capital is a function of many factors. The technology a firm employs to produce and deliver products depends, among other things, on how many people have access to it and whether they are willing to use it. These considerations are even more important for high-contact industries, that is, industries in which there is a frequent interaction with the clients and clients access the service several times a week or even a day. For example, telephone banking would have little rationale in a market or age in which few households had an easy access to a telephone set. Similarly, the success of the automated banking machines (ATMs) is closely related to their availability in accessible locations and sufficient numbers. The existence of a critical mass of clients using the technology as the point of access is of primary concern for the commercial viability of investment in it. Therefore, the task of gathering intelligence on connectedness is multi-dimensional. Data on technology adoption by business and on households' access to the Internet are both essential to understand the impact on the economy. The early work has focused on the household sector, providing data on such indicators of connectivity as the ownership rates of and acces to personal computers and modems, connections to the Internet and its major uses (Dickinson and Sciadas, 1999).

This study is a preliminary effort to address the other side of the equation: the diffusion of Internet technology in the business sector. The focus is on selected industries in the financial services sector, specifically banks, trust companies, life insurers and property and casualty insurance companies. The financial services sector includes other industries as well but only these four were covered in the survey. Nevertheless, for the sake of brevity we will refer to these industries as the financial services sector or the banking and insurance industry.

3. DATA AND DEFINITIONS

As is the case with any new concept, defining electronic commerce and developing a measurable definition is an evolving process. The concept and content of electronic commerce as well as the channel over which it is conducted are currently the subject of discussion. Some definitions of electronic commerce are available in the OECD (1998) and Government (1998). Further, from the point of view of measurement, it is important to ascertain how firms interpret these terms and whether firms within an industry and across industries have a uniform view. Lessons learned from the early measurement efforts such as this one and the analysis of data collected should provide useful inputs and advance the ongoing discussion, leading to a convergence of views on what should be measured and what is measurable. Given that there is a choice of definitions, the following paragraphs clarify what precisely is measured for the present study. The purpose is not to advance new definitions but rather to provide the context for interpreting the data.

The analysis is based on statistics gathered as part of the Survey of Innovation (1996), the first results of which were reported on in Baldwin et al. (1998). The following is a brief description of the relevant portions of the survey; detailed methodology and its implications for data quality are discussed in Hamdani (1999). A set of questions on the acquisition of the Internet technology and selected uses was included in the questionnaire as a prelude to more detailed surveys of electronic commerce as well as to gather data to shed light on the relationship between the application of information technology and firms' ability to innovate. Since it was not a survey of information technologies the questions were kept as simple as possible.

The question on electronic commerce was quite specific with respect to the technology or channel to be included and less specific with respect to the activities. As the Internet is the latest technology, our interest was to know to what extent this new channel was being used to conduct commerce. All other electronic channels were excluded. Electronic technologies in the financial services industry are commonly understood to include telephone banking, transactions conducted by automated banking machines, debit machines or point of sale terminals and so on. In order to avoid confusion between the Internet and other electronic technologies, questions were designed to first focus respondents' attention on the Internet and then lead up to the questions on applications. Although the definition provided in the questionnaire was specific as to the delivery channel it did not state that the Internet technology meant only the open Internet structure, and transactions conducted on the PC using proprietary systems should not be included.

As to the electronic commerce activities, firms were asked whether they used the Internet to sell their products. This would exclude firms that conducted electronic business including purchases but did not sell on-line. The word electronic commerce was not used because it was not clear whether all or most firms had a common understanding of the term.

The survey was conducted from February to May 1997 and requested information for 1996. It is likely that in some cases data pertain to the period during which the questionnaire was being completed rather than the reference period for which the information was requested. The response rate was about 70 per cent with some variations from industry to industry. There is likely to be an upward bias in estimates weighted by operating revenue for several reasons including a significant amount of restructuring and consolidation going on in the industry since the de-regulation began and particularly in recent years. However, the conclusion that the estimates of the pervasiveness of Internet technology and its various uses based on revenue are significantly higher than those based on a simple count of firms is not affected.

4. ADAPTABILITY OF FINANCIAL PRODUCTS TO THE INTERNET

Some products are more amenable to transactions on the Internet than others and some industries are better prepared or under more pressure from competitors to adopt this technology. First, intangible products are more readily accepted by the buyer than the tangible products. As many services are produced and consumed simultaneously, the buyer relies on his experience and the provider's reputation to judge the quality of the product. Second, transactions in intangible products are more convenient to execute. The delivery is instantaneous for at least some intangible products because the point of sale and point of delivery is the same. Transactions for tangible products, on the other hand, involve two different technologies -- communication and information technology to make the commitment to exchange the product for money, and transportation technology to deliver the product. In other words, the buyer places the order on-line and the supplier delivers it by mail or courier service. Financial intermediary services, music and software are among the prime examples of the products for which the point of sale and point of delivery is the same. Further, in a market where trust and security matter a lot, shoppers on the Internet are more likely to deal with reputable companies or products with known brand names. Financial intermediaries fit this description. With trademarks as symbols of quality and with the advantage of product identification through recognizable brand names, financial products are well suited for transaction on the Internet. Finally, among products for which estimates are available, on-line transactions in banking and insurance products are among the most economical. Purchasing a term life insurance policy on-line costs one-half and an average retail banking transaction one-tenth the cost incurred with the traditional technologies (Colecchia, 1999).

Product adaptability to digital technology is only one side of the picture. Just as important is the industry's ability to offer products on-line. Electronic transactions are not new to the financial services industry. It has been at the forefront of new technologies for decades beginning with the automation of backroom operations and the subsequent introduction of electronic fund transfer. Its experience in dealing with clients in retail business is well established, with the introduction decades ago of automated banking machines followed by telephone banking and PC banking. Indeed, the financial services industry is one of the main producers of the information technology, second only to the information technology industry in North America (Darlington, 1998).

5. ESTIMATES OF NEW INDICATORS OF INFORMATION SOCIETY

Technological diffusion can be measured from different angles, depending on the purpose of the analysis. One can simply count the number of firms that have adopted the technology. This type of statistics is primarily useful for informing science policy discussion because it indicates the technological advancement of the society. Another indicator is the value of economic activity of the firms using the technology as a fraction of the total activity of all firms in the industry; in technical terms, it means producing estimates weighted by an economic variable. This indicator is useful from an economic point of view because it measures the size of economic activity that could potentially be affected by the technology. Since it is a measure of the total activity of connected firms it does not take into account the fact that firms may be using several technologies at the same time. A more precise measure would be data on only that portion of activity that is due to the Internet, that is, the amount of economic activity transacted on the Internet. This is the ultimate metric of the electronic commerce and the eventual goal of measurement efforts.

5.1 Diffusion of the Internet Technology

In 1996, more than half (56.1 per cent) of the financial intermediaries had adopted the Internet technology (Table 1). The average for the sector is driven down by the property and casualty insurance industry that has a relatively high weight in terms of the number of firms and a low Internet adoption rate. Proliferation rates vary significantly among individual industries. Internet technology was more pervasive in the banking industry with an adoption rate of 73 per cent than in any other financial services industry. Two-thirds of the life insurance companies (66 per cent) reported using the Internet and trust compnaies followed closely behind with an adoption rate of 63 per cent, while fewer than one-half (45 per cent) of the property and casualty insurance companies were linked to the Internet in 1996.

Table 1. Internet Proliferation Rates by Firm Size, 1996 (%)

	All firms	Firm Size (Operating revenue)		
		Above $100m	**$25m - $100m**	**Under $25m**
Internet Proliferation rate	56.1	82.1	59.2	37.8
Internet applications: (% of Internet users)				
-E-mail	70.8	75.0	67.2	62.7
-Web searches	87.2	93.7	83.6	90.4
-Internet commerce	17.0	19.5	17.9	12.2

Source: Statistics Canada (Science, Innovation and Electronic Information Division), *Survey of Innovation 1996.*

Comparable data to place the connectedness of the financial intermediation industry in perspective are not available. The Survey from which data for the present paper was collected covered only selected industries in the service sector and therefore did not provide economy-wide estimates of the Internet proliferation rates. However, various private sector estimates place this industry at the forefront both of the adoption of the Internet and of conducting electronic commerce. Regular quarterly surveys of member organizations conducted by the Canadian Federation of Independent Businesses provide a context. In the fourth quarter of 1996, 40 per cent of the larger members were connected to the Internet while the figure for small firms (employing under 5 employees) was one-half that for large firms (Mallett, 1999).

While the digital strategies of firms depend upon a number of factors including the nature of business, investment in the legacy systems and the cost of new technology, competitive environment, intensity of the competition, etc, larger firms are more likely to have the Internet technology than their smaller counterparts. For example, 82.1 per cent of the firms with operating revenue above $100 million reported using the Internet in 1996. The adoption rate dropped to 59.2 per cent for firms in the operating revenue range of $25 million to $100 million, while only 37.8 per cent of the small firms (operating revenue under $25 million) were linked to the Internet. However, as we shall see later, small firms make more intensive use of the Internet than their larger counterparts.

The simple count of firms is a useful indicator of technology diffusion in the economy from the point of view of the science and technology policy. However, as an indicator of the impact of Internet technology on the economic and business life of the country, it is inadequate because it treats all firms in an industry and all industries in the sector as equal, regardless of their size. Firms with a larger client base and higher volume of business have a greater impact on the economy than the smaller firms, and the same applies to industries. Therefore, an indicator that takes into account the size of the firm in terms of economic activity provides a more accurate picture of technology diffusion through the economy. Operating revenue and the number of employees are the two most common measures of firm size. Operating revenue is preferable as employment is a less relevant measure of the size because of the substitutability between technology and human resources, growing use of contract and temporary workers and consultancy services -- some of whom are not necessarily defined and counted as the labour input of the firm in which they perform work under the classification systems -- and outsourcing of non-core activities. Moreover, some types of financial intermediation operations are more labour-intensive than others. For example, retail financial operations may require more employees to generate the same amount of fees than corporate or investment banking.

When the economic weight of firms is taken into account, the results are considerably different. The influence of the Internet is much more pervasive in the economy than technology adoption rates based on a simple count of firms indicate. The proliferation of the Internet technology measured on this basis is higher for the sector as a whole as well as for each individual industry. Firms connected to the Internet accounted for 90.9 per cent of the financial services sector's operating revenue in 1996 (Table 2). As explained above, this figure likely has an upward bias. The actual estimate would be somewhat lower but still considerably higher than that based on a simple count of firms.

Table 2. Innovation Rates, 1996 (%)

	Firms using the Internet	**Firms not using the Internet**
Overall Innovation Rates	74.2	35.8
Frequency of Innovation:		
More than three times a year	20.8	6.1
Three times a year	6.2	4.5
Twice a year	17.4	16.7
Once a year	25.8	30.3
Every second year	12.9	1.5
Every third year	1.1	15.2
More seldom	15.7	25.8

Source: Statistics Canada (Science, Information and Electronic Information Division), *Survey of Innovation 1996.*
* See text, section on "Data and Definitions" for data quality

5.2 Business-related Uses of the Internet

In the discussion of electronic commerce, a distinction is drawn between electronic commerce and electronic business. While the two concepts and their measurable definitions are evolving, electronic business is a broader concept and includes electronic commerce as a sub-set. Several business-related activities may take place before a transaction (electronic commerce) occurs. These activities may include gathering of information on prices and the quality of products (goods or services), browsing to narrow down the selection and finally a decision to purchase a product before consummating the transaction. Two of the most common, easily identifiable and easy to measure applications of the Internet are e-mail and searches on the World Wide Web (WWW).

Searching on the Web is a more common activity than e-mail. Among firms that reported using the Internet, 87.2 per cent (Table 1) did searches while a smaller number, 70.8 per cent, used it for e-mail. This pattern was observed across all industries -- chartered banks, trust companies, life insurers and the property and casualty insurance companies. The use of e-mail to conduct business can be constrained because of the privacy concerns of the parties involved. Further, firms operating in niche markets and dealing with only large clients may prefer to discuss and negotiate business in person or by tele-conferencing rather than e-mail.

It is interesting that while the size of firms is closely related to the Internet adoption rate and its various uses (e-mail and electronic commerce) this is not the case with respect to the use of Internet technology to do Web searches. Small firms (earning under $25 million) are more likely to do Web searches than their larger counterparts. In 1996, 90.4 per cent of the small connected firms used the Internet for that purpose, well above the figure for medium-sized firms (83.6 per cent) and very close to the 93.7 per cent for large firms (Table 1). A possible explanation is that while large firms may have other sources of information at their disposal, small firms try to maximize their investment in Internet technology and use it as the principal source of gathering information and analyzing it for use in their business strategy.

5.3 Internet Commerce

The use of the Internet to sell products on-line is less common than for other purposes discussed above. Seventeen per cent of the firms (and 59.5 per cent when weighted by operating revenue) which were connected to the Internet reported selling products on-line. Selling on-line was least common among the property and casualty insurance companies with only 9 per cent of the firms engaging in this activity and most common among the trust companies with a rate of 34 per cent. Banks and life insurance companies were within this band with 23 per cent and 17 per cent respectively.

While firms of various sizes sell on-line, Internet commerce seems to be very much an activity of the large firms. As with proliferation rates and e-mail, the number of firms engaged in Internet commerce declines with the size (Table 1) -- 19.5 per cent of the large firms, 17.9 per cent of the medium-size firms and 12.2 per cent of the small firms reported selling on-line in 1996. The importance of large firms is better conveyed by comparing their share of connected firms with their share of firms selling on-line. Organizations earning more than $100 million in operating revenues in 1996, accounted for just over one-quarter of all firms but about one-half of the firms that used the internet to sell their products.

Chart 1: Technology and Efficiency

Cost of An Average Retail Banking Transaction

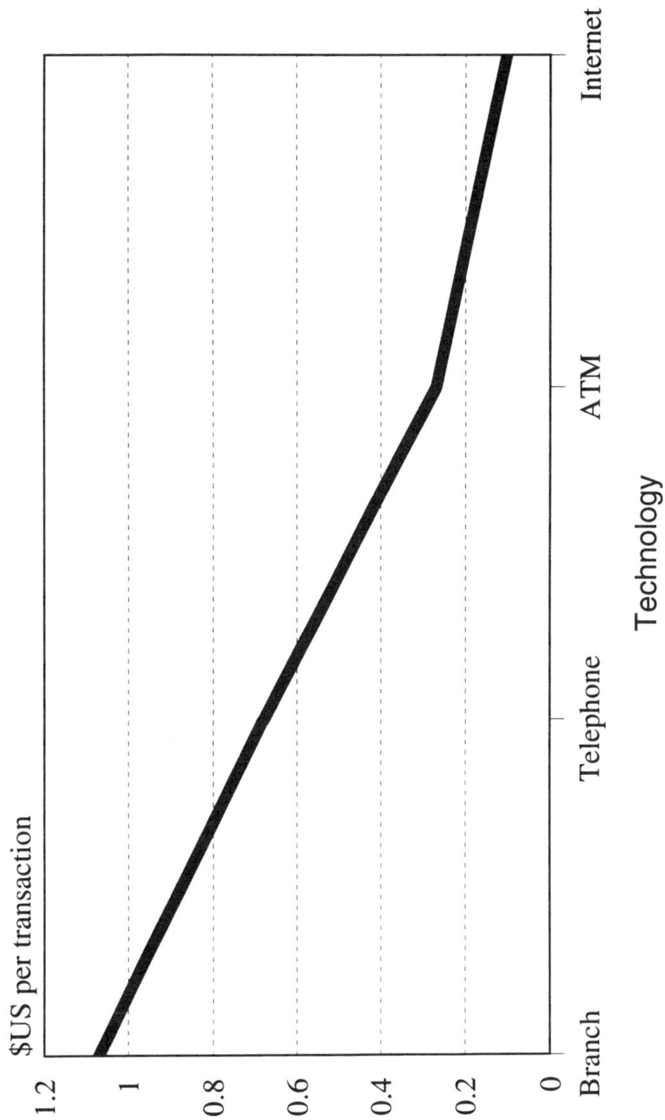

Source: Larry Downes and Chunka Mui (1998)

In 1996, electronic commerce was only beginning to attract attention. Only 7.4 per cent of the Canadians households (844 thousand) were linked to the Internet although this figure is increasing rapidly and reached 13.0 per cent in 1997. More importantly, only 3.1 per cent of the households did electronic banking (Dickinson and Sciadas, 1999) of which the Internet banking is only a part. In spite of the growing popularity of electronic banking and its emergence as the main delivery channel for retail transactions, it constituted only a small portion of the total services provided by deposit-taking institutions in 1996 (Statistics Canada, 1998). Since, the large firms offer a wide choice of delivery channels to their clientele to suit their individual preferences the volume of products sold on the Net was small compared to their total revenue in 1996. As the number of products sold on the Net increases and the number of clients connected to the Internet grows, Internet commerce will rise. The incentive for both the financial intermediaries and clients to use the Internet is evident from the convenience and low cost (Chart 1).

6. EMPLOYEE ACCESS TO THE INTERNET

The discussion has so far centred on the pervasiveness of Internet technology among firms. Proliferation rates within firms provide an indication of how the workplace is changing as well as a measure of the intensity with which firms use the technology. An appropriate measure of intensity would be the number of transactions conducted on the Internet or better still the volume of commerce. However, this raises the difficult questions of the concept and a measurable definition of electronic commerce and electronic business. Our understanding of what these terms mean and whether the respondents who provide this information have the same understanding as those who gather the intelligence is still evolving. Hence, we used a proxy and asked respondents to indicate what proportion of their employees had access to the Internet. The access was defined to include availability from the employees' desk. For the purpose of the following analysis, intensity is taken to be higher if a larger percentage of the firm's employees has access to the Internet from their desks and lower if fewer employees can access it.

In 1996, nearly two-thirds of the firms reported that fewer than 10 per cent of their employees had access to the Internet from their desks. It was available to all employees in only 11.6 per cent of the firms (Table 3).

Table 3. Employee Access to the Internet within Firms by Firm Size, 1996 (%)

	All firms	**Firm Size (Operating revenue)**		
		Above $100m	**$25m - $100m**	**Under $25m**
All employees	11.6	7.9	10.4	17.9
75% - 99%	3.3	2.0	3.0	5.4
50% - 74%	2.9	3.0	3.0	2.7
25% - 49%	7.9	6.9	4.5	12.2
10% - 24%	9.9	7.9	17.9	5.4
Under 10%	64.3	72.2	61.2	56.3

Source: Statistics Canada (Science, Innovation and Electronic Information Division), *Survey of Innovation 1996.*

The access to technology is a function of several economic decisions including the nature of business and a firm's desire to maintain an optimal labour/technology ratio. The ratio tends to be higher (and access lower) in firms which are diversified and operate a network of branches. In smaller firms, on the other hand, which serve niche markets and do not operate a large network of branches the ratio is lower (and access higher).

Unlike the relationship between firm size and Internet adoption rates among firms, the proliferation rate within firms moves inversely with the size. It implies that small firms make more intensive use of the Internet than the larger firms, using the above definition of intensity. Among firms with operating revenue above $100 million, only 7.9 per cent reported that all their employees had access to the Internet from their desks. It rose to 10.4 per cent for the mid-sized firms and jumped to 17.9 per cent for small firms (Table 3). The same conclusion held when the relationship between firm size and intensity was analyzed by looking at the percentage of firms where the least number of employees had access to the Internet. Nearly three-fourths of the large firms as compared to just over one-half of the small firms reported that less than 10 per cent of their employees had access to the Internet from their desks.

6.1 Digital Strategies of Financial Intermediaries

Many factors enter into a firm's decision to invest in a new technology, which determines the number of firms that will be linked to the Internet and offer products on-line. They include its age; its perception of how the competitors are responding to technological advances; the client base; clients' preferences; the cost of technology already in place and when it was installed; the cost and benefits of the new technology; and the firm's forecast of technological change. Broadly, three types of strategies appear to emerge with respect to the Internet technology. Some firms are entirely driven by the digital strategy. Business strategy is more important for others. In-between, there are firms that employ a mix of business and digital strategies.

A new startup firm is more likely to have the latest technology and engage in Internet commerce. New firms often enter the market enticed by opportunities opened up by the emergence of a new technology. In order to succeed in a market in which big players already have a very large share and existing small firms have carved out niches for themselves, a new entrant has to prove why the country (or world) all of a sudden needs it when it has been getting along without it for years. The new technology often provides it with the tool to do just that. It reduces the cost of production or delivery, enables the firm to introduce new products not previously available, improve the quality of existing products and make access easier and more convenient. The remarkable reduction in cost achieved by the introduction of successive technologies in the banking industry is summarized in Chart 1. These data are based on the U.S. experience but are also used in the Canadian context to indicate the trends. There was a sharp drop in the cost of an average retail banking transaction when the telephone banking was introduced, and the use of automated banking machines made it still more efficient. Using the Internet, the cost is less than one-tenth that of an in-branch transaction and is declining further, not to mention the convenience of being able to do business at any time and from anywhere.

In addition, the cost of entering the market has declined with the emergence of Internet technology relative to what it was with the traditional technology -- branch network. The cost of setting up a fully functioning Web site for a full service retail bank is estimated to be less than opening a brick-and-mortar branch office. Add to that the fact that the Web site of a virtual bank is accessible to all households with Internet connection in the country and beyond, while the reach of a traditional branch is confined to a limited market bounded by geography.[1]

Benefits of new technology notwithstanding, virtual corporations are a recent phenomenon, and were rare in the period covered by this study. In addition, the rates of entry into the financial services industry which determine the number of new firms that will enter the market and use the latest technology are low compared to other industries (Hamdani, 1997 and 1998), partly because of the regulatory controls. Therefore, the impact of new firms as a determinant of Internet proliferation was very small.

On the other end of the spectrum are those existing firms that did not have an Internet connection. Possible explanations include the nature of their business, the type of market they serve and their clientele. A firm with a retail type of operations, for example, is more likely to have an Internet connection and sell on-line than one serving a niche market of a few, large business clients with whom it would like to maintain a person-to-person to contact.

In-between these two extremes lie a large number of firms that use hybrid technologies including the Internet. They are not just driven by opportunities offered by the Internet but are also guided by a business strategy based on the fact that they are dealing in a customers' market with diversified preferences as to the technology, the cyberspace marketplace is global and the competition comes not just from firms classified to the financial services industry by the Industrial Classification Systems but also from others including intelligent agents and integrators.[2] Unlike new firms that seek market niches and concentrate on clients who prefer to use the Internet as the point of access, existing firms satisfy the demands of a diverse clientele. Many of their existing clients do not have access to the Internet, it may not be the preferred choice of some, while others are used to alternative technologies. However, they are aware that the new marketplace is a customers' market and, just as the needs of those who prefer to transact business in-branch must be met, there is a growing segment of the population that is linked to the Internet and prefers the convenience of doing business on-line from anywhere and any time.

For these institutions striking a balance between offering services in-branch and on-line on the one hand and competing with low cost competitors who only offer electronic delivery channels on the other becomes crucial. It is therefore not surprising to see an increasing number of strategic alliances among financial institutions to achieve the economies of scale, alliances with retail businesses to set up financial services kiosk to expand the client base and alliances with software companies to compete with intelligent agents and integrators.

7. INFORMATION TECHNOLOGIES AND INNOVATION

The speed and low cost of information made possible by the Internet has placed enormous power in the hands of the clients and given financial institutions vast opportunities to innovate and create value for their customers. With the help of the software and other Internet-based software companies, customers can shop and go for the best deal foregoing popular brand names. Just as the customers put information to their advantage, so too do the financial institutions. With the advent of the Internet, nearly every transaction consumers make leaves vast information about their product preferences, their purchasing power and their spending and investment patterns at the disposal of companies. Leveraging information has become a key ingredient into the value these institutions create for their customers in order to retain their business and loyalty.

The micro-segmentation of the market -- a high degree of product customization -- made possible by the emergence of the Internet leads to shorter product life cycles and shorter business planning horizons -- estimated to be one year or even less as compared with the usual five-year plan (Downes and Mui, 1998). We try to shed light on two questions: Is the Internet an enabling technology for innovation and has it shortened the product life cycles? Are users of the Internet more innovative than non-users and do they innovate more frequently? The definitions of innovation employed in the analysis are the same as in the Oslo Manual (OECD, 1997)

The evidence from the survey data appears to show that the answer to both questions yes. Firms using the Internet technology are more innovative than the firms that are not linked. The differences are remarkable. In 1996, 74.2 per cent of the firms that had Internet technology reported introducing a new product or process; the comparable rate for firms that did not use the technology was less than one-half, at 35.4 per cent (Table 4).

Table 4. Innovation Rates and Frequency of Intervention,* 1996 (%)

	Firms using the Internet	**Firms not using the Internet**
Overall Innovation Rates	74.2	35.8
Frequency of Innovation:		
More than three times a year	20.8	6.1
Three times a year	6.2	4.5
Twice a year	17.4	16.7
Once a year	25.8	30.3
Every second year	12.9	1.5
Every third year	1.1	15.2
More seldom	15.7	25.8

Source: Statistics Canada (Science, Information and Electronic Information Division), *Survey of Innovation 1996*.

* Innovation is defined to include both product and process innovations.

Further, the frequency with which firms brought new products and processes to the market differed considerably between the users and non-users of the Internet. Over one-fifth of the Net-users introduced more than three innovations a year during the reference period compared with only 6.1 per cent of the firms not using the Internet. Another way to examine the frequency is to look at firms that seldom innovated. Here again, the Internet users fared better. Just under 16 per cent of them reported that they innovated less often than every third year while nearly 26 per cent of the firms that did not use the internet technology took more than three years to introduce a new product or process.

8. CONCLUSIONS

This paper presented the results of the first effort at Statistics Canada to measure the pervasiveness of Internet technology in Canadian business and its various applications including the Internet commerce. Over one-half of the financial intermediaries were linked to the Internet in 1996 but the figure was much higher for the banks at 73 per cent. The most common use of the Internet in 1996 was for Web searches followed by e-mail, while only 17 per cent of the firms reported selling products on the Internet. However, the Internet technology is more pervasive when differences in firm sizes are recognized and the results are weighted by an economic variable. The data reveal that Internet proliferation rates and its various applications decline directly with the firm size. However, smaller firms make more intensive use of it, partly because of the nature of business and the desire to make an optimal use of their limited resources. They are also likely to rely more on the Internet to gather information, absorb it and leverage it to create value for their customers than their larger counterparts who may have access to other sources such as their own research facilities or agreements with other firms.

The results of this first effort appear promising. The response rates for questions on the Internet and its various uses were robust, and it was possible to determine the direction if not the magnitude of biases in the indicators constructed.

The use of the Internet is spreading at a very rapid rate, and the data become dated just as fast. The results presented here serve as useful benchmarks. As more comprehensive surveys are undertaken in the future, we can look back and see how far along we have come and where we stood in 1996.

Notes

[1] In 1997, the startup cost of a Web site was estimated at $1 million, about the same as that of opening a branch (excluding the cost of real estate). However, to reach the same customer base by branch network, an investment of $900 million will be required (Darlington, 1998).

[2] Intelligent agents refer to Internet-based value added chains or software that shop for the best value available -- such as the highest term deposit rate or best premium for insurnace -- on behalf of their clients, and "integrators" manage finances on behalf of their clients. These are usually software specialists who by offering these conveniences threaten to become intermediaries between the financial institutions and their customers.

Chapter 8

Innovation, Information Technologies and Human Capital in the German Service Sector

GEORG LICHT
Zentrum für Europäische Wirtschaftsforschung (ZEW), Mannheim

1. INTRODUCTION[1]

In the last decade most industrialized countries experienced a steady growth of value added and employment in the service sector. The employment growth of the services was a major engine for additional jobs, which helped to overcome the decrease in the number of jobs in manufacturing. Given the historically high level of unemployment in Germany today, stimulating the development of the service sector is of great concern to policy makers. In addition, international comparisons reveal that the employment share of the service sector in Germany still accounts for nothing like the proportion of jobs it currently provides in the USA or Canada, for example.[2] In 1994, almost two-thirds of Germany's gainfully employed population were working in the service sector (including public-sector employees). This means that its ratio to the other sectors has reversed within just 35 years. This ongoing transformation into a service society becomes all the clearer when we also examine the changes in activity structures throughout the manufacturing sector, where for years now we have been seeing an accelerating shift away from traditional production operations and towards more service-oriented activities.

Moreover, there is also some concern about how these changes of the structure of our economies will affect productivity and hence the ability to sustain high wages and a high living standard. As shown by recent research, long run productivity growth in services seems to be slower in most service industries than in manufacturing (see e.g. van Ark, Monnikhof and Mulder 1999 and O'Mahony 1999). Although this partly can be attributed to measurement error, this is certainly not the whole story.[3] However, in recent years productivity growth in services has increased. This increase can at least partly be attributed to the increasing importance of product and process innovation in services. The diffusion of information and telecommunication technologies and the regulation in services have been the driving forces behind this development. Increasing innovative activities in services as well as the increasing importance of service innovation for the economy as a whole will certainly affect the employment prospects of this sector. And, it is an open question whether the expansion of services in the future will maintain its role of the job engine it had in the past. Especially as increased technology use in services will probably change the skill structure in services in the same way as it is well known from past developments in manufacturing. So, the expansion of services will probably not help to overcome the pressing structural problem of unemployment which is manifested in much larger unemployment rates for low skilled workers relative to skilled workers in nearly all OECD economies (see e.g. Legler *et al* 1999).

This paper addresses these problems. In the first part we will introduce the data used for the study and discuss some of the associated measurement problems of innovative activity and the use of new technologies (especially IT) in services. In section 3 we will discuss selected results of the German innovation survey in services. We will highlight the share of innovative businesses in selected service sectors, the importance of various technologies for innovation in services, and the main sources of service sector innovations. Section 4 deals with implications of innovation for the structure of labor demand of service sector businesses. Finally section 5 summarizes the discussion and provides some consequences for innovation and education policy in Germany.

2. DATA AND MEASUREMENT ISSUES

Our study is based on the first and the second wave of the Mannheim Innovation Panel for the Service Sector (MIP-S). The first wave develops a concept for surveying innovation in services by extending the well-known instrument of innovation surveys in manufacturing to the service sector. The second wave is based on the second edition of the OSLO manual, which explicitly considers service sector innovations (OECD 1997). Both waves employ mail surveys to generate the data. The first wave took place in 1995-1996.[4] The second wave, which also represents the German part of the community innovation surveys (CIS), coordinated by EUROSTAT, was conducted in 1998.

Topics covered in the questionnaire include (1) general information about the firm (size, industry, sales, number of employees, labor costs, exports, strategic management objectives, customers and product characteristics), (2) the human capital of the firm and (3) investment in new physical assets and in information technologies. Special attention was paid to the innovative activities of firms with questions on R&D personnel, the acquisition of knowledge, co-operative innovation activities, linkages to public R&D institutions, on the importance of various new technologies and on the impacts of innovation. In addition to the EUROSTAT's harmonised questionnaire the German survey also includes questions relating to the most important technologies used in connection with product and process innovations and firm's spending on vocational education and training.

The survey's population consists of all service sector firms with more than four employees, contained in the records of CREDITREFORM, which is Germany's largest credit-rating agency.[5] This source provides data on the number of employees, industry affiliation and the addresses of firms. This information was used to stratify the sample by sector and firm size classes, as well as to identify whether the business was in the former West or East Germany. The MIP-S is restricted to marketed services and, therefore, only includes firms from wholesale and retail trade, transport, banking, insurance, software, technical consultants, marketing and other business related services.[6]

The overall response rate in both waves was around 25%, which seems fairly reasonable when compared to response rates for other voluntary mail surveys in Germany. The response rate was above average for the firms in the banking sector, and for the firms in technological services and software firms and it was below average for the firms in wholesale and retail trade. Participation slightly increased with firm size. In addition to the mail survey a short telephone survey of non-respondents was conducted comprising in both waves another 1,000 firms in order to uncover selection effects correlated with innovation activities. Comparing the results from the mail and the telephone survey we conclude that there seems to be only a small response bias in favor of innovative companies (see Licht 1997 et al. for details). Therefore, the share of innovative companies especially in the traditional service sector, retail and wholesale, trade and transport services, is probably slightly upward biased.[7]

Overall, our experience with innovation surveys in services turned out to be quite encouraging. The definition of innovation in the service sector is more or less similar to that used in the manufacturing sector: i.e. the introduction of new products and services or the implementation of improved processes. Some important measurement issues, however, remain yet to be resolved. These issues comprise the measurement of the output of innovations as well as the measurement of some innovation inputs.[8]

Expenditure on research and (experimental) development as well as the number of R&D employees are the traditional measures of the innovative effort. A vast number of studies on the economy-wide level or the level of single industry show that R&D expenditures are correlated with innovation output e.g. productivity growth. In services, some problems arise from the definition and interpretation of research and development (R&D). The traditional concept of R&D measurement in Germany, based on OECDs Frascati Manual (OECD 1993), is mainly related to technological innovations in the manufacturing sector. Consequently the definition of R&D in the service sector sometimes is biased against service sector innovation activities. The partly immaterial character of services (e.g. service innovations do not relate to the process of transforming physical goods but improving the provision of information) and the interactive nature of innovation in business services complicates the measurement problem. As a consequence the level of R&D expenditure in the service sector based on R&D surveys will probably be strongly dependent on the questionnaire design and the way R&D is defined. The number of R&D performing firms and their R&D expenditures is therefore quite sensitive to methodological changes in the data gathering process.

This is exactly what we experienced. We found that the interpretation of R&D by service sector firms is - at least in some cases - not compatible with the definition of R&D in official R&D surveys. Furthermore, using the official German R&D figures as a benchmark we conclude that the role of R&D activities in German service sector companies is underestimated. However, involvement of a firm in R&D activities (as stated by the firms) can be used as an indicator that a firm devotes resources to product development and design activities.

Measurement problems of the innovative output in services are also related to the interactive nature of many service innovations. Furthermore, productivity measurement problems in services arise due to the use of new technologies because, unlike the manufacturing case, innovation in services often is neither represented by new services nor by process improvements, which increase output or decrease input requirements. Innovation in service is often more closely connected to the way products are delivered. For example, technological change is often associated with the number of hours during which a service can be delivered or with improvements in the spatial dimension of the services (e.g. home banking). The quality of a service (e.g.

user-friendliness) is another component of innovation which is usually not correctly reflected in the measurable output (see Licht and Moch 1999, or Baldwin 1998 for an approach to measuring service outputs in qualitative terms). The multidimensionality of service sector innovation output partly explains the existing problems in tracking productivity growth of services. Lower productivity growth figures of service industries, relative to manufacturing, should be regarded with care. Improvements in service production and qualitative improvements of service products have an impact on services and the whole economy. These improvements will be one driving force of structural change, generating incomes and probably additional jobs although the full effects are probably not reflected in the measured product of services but in the product of the customers of service firms.

Another conclusion can be derived from relating innovation output and potential inputs to innovation processes of service sector firms. To do this we shortly highlight the results presented in Licht and Moch (1999) who view IT-investment, R&D, human capital and physical investment as main inputs in innovation in services. In summary, they found significant effects of IT-investment on flexibility in adjusting products to customer needs, user-friendliness of the products, the temporal availability of services, and delivery speed, but only a weak impact of IT-investments per employee on the productivity of the firm. Thus, we conclude that the effects of IT-investment on innovation in services are reflected mainly in product quality improvements. Direct productivity impacts or quality improvements of the production process for services seem to be less affected by IT. Non-IT physical investment (calculated as total investment minus IT-investment) is especially important when the flexibility in adjusting products to specific customer needs, fulfilling legal standards and regulation, and the productivity of customers are viewed as innovation output of services. Finally, R&D efforts seem to be relevant in enhancing the productivity of the firms at hand as well as the productivity of customers. R&D is also associated with components like reliability of the service and the ability to fulfill various regulatory requirements. Similarly, firms' expenditures on human capital enhance productivity and the quality of services.

Human capital investment turned out to be even more important in services than in manufacturing. Hence, an innovation survey in services should try to measure human capital input, which can be thought of as playing as crucial a role for service innovations as R&D does with regard to manufacturing innovations. Moreover, human capital is also a key enabler for the diffusion of IT in services. To realize the growth opportunities generated by information technology in modern service industries, human capital plays a crucial role. Therefore, we should be more concerned with the quality and quantity of the output of our educational systems in the knowledge-based economy than in the past.

3. INNOVATION AND INFORMATION TECHNOLOGY IN SERVICES

The service sector should no longer be accused of failing to innovate. Deregulation in many areas of the service sector and the rapid development and diffusion of IT technologies opened up the market for international competition. Manufacturing companies concentrate on their core competencies and increasingly source out ancillary service activities. Taken together these developments increase competition in services and open up additional opportunities for companies to improve their market positions by introducing new products, and to enhance their efficiency by introducing and reinventing their production processes. We, therefore, should expect to find the largest share of innovating companies[9] in those sectors where market conditions have been undergoing the most intense changes. As shown by Figure 1 this expectation is confirmed by our data set.

The largest share of innovating companies is found in the telecom sector which was deregulated in Germany in the mid-nineties. This opens up new market opportunities which can be realized by new product introduction and put a pressure on the previous monopoly firm to increase the productivity. Following in the row software and IT service firms. Given the large rapid technological development e.g. introduction of intranet and internet this also comes not as a surprise. However, somewhat surprising engineering and technological business services comes next. Given the slacking demand in manufacturing and in the construction sector at the time period covered by our survey there is a large pressure to increase the efficiency by process innovation to cope with falling prices in that period. As is indicated by Table 1 the introduction of IT technology plays a central role in this sector.

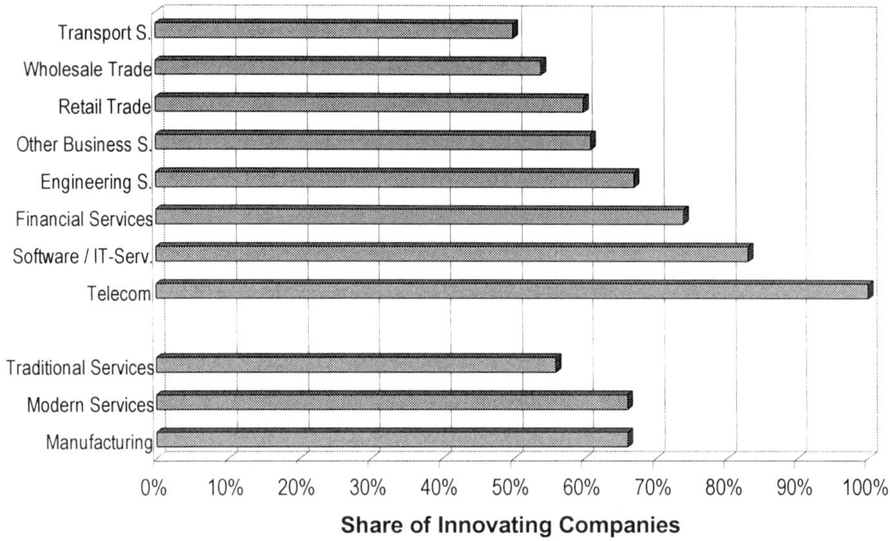

Figure 1. Share of Innovating Companies in Selected German Service Sectors in 1996

Source: Mannheim Innovation Panel in Services (MIP-S) 1998

Adding the other business services which consist of a rather heterogeneous set of firms these sector can be comprised using label 'modern service sector'. In average the share of innovating firms in this group of industries is quite similar to the average share of innovating companies in the German manufacturing and mining sector[10]. Even this simple comparison shows that innovation is no longer the domain of manufacturing. New technologies change services in increasingly similar ways they chanced manufacturing in the past.

Innovations in manufacturing and in services often are interrelated. The service sector aids innovation in manufacturing by providing know-how. Service companies on the other hand are major of customers and, therefore, important sources of knowledge for manufacturing based innovations. This is obvious from Table 1. Table 1 provides data on the share of companies, which have invested in new physical investment, the share of companies, which invest in information technologies as well as the share of companies, which provided further vocational education and training to their employees.

First of all, the overwhelming majority of companies in the surveyed service industries invest in information technology. Of course, in the modern service sectors is large than in the traditional sector. However, the mere size of the traditional services sector make them to important customers of information technology and as such strongly influence the development of this technology. As it is, well-known financial services are one of the largest spender of IT. In addition, we also note that in the modern services investment in new physical good is closely related to the investment in IT. The share of firms with investment in new capital goods is more or less identical with the share of firm investing in IT in these sectors.

Table 1. Innovation, Investment and Expenditure on Education

Share of Firms (1996)	**Innovative Firms**	**Firms buying capital goods**	**Firms with investment in IT**	**Firms expenditure on CET and CVT**
Wholesale Trade	54%	84%	79%	62%
Retail Trade	60%	92%	82%	73%
Transport	50%	88%	74%	51%
Telecom	100%	100%	100%	100%
Banking/Insurance	74%	91%	89%	90%
Software, IT services	83%	97%	97%	88%
Engineering Services	67%	100%	98%	81%
Other Business Services	61%	94%	88%	76%

CET = continuing vocational education; CVT – continuing vocational training
Source: MIP-S / CIS II – Germany

However, IT is not only the most important investment good. It also plays a central role for innovation in the service sector. Table 2 contains the share of innovating companies which state that based on list of given technologies.[11] The dominant role of information technologies for innovation in services becomes immediately clear when look at this table. Even in the least IT-active sector – retail trade – 87% of the innovators believe that IT was important for the innovations introduced. The importance of all other technologies vary vastly by sector. New facility management techniques are important to engineers e.g. for drafting blue prints of new buildings. Latest transport technologies are – of course – crucial to the transportation sector.

Table 2. Use of Different Technology for Services Sector Innovations by Industry

As share of innovating companies in 1996	Information	Environ -mental	Facility Management	Transport	Measure ment and Control	Life Sciences	Nutrition	New materials	Others
Wholesale Trade	88%	45%	27%	49%	22%	6%	5%	21%	4%
Retail Trade	87%	51%	5%	25%	12%	4%	1%	10%	4%
Transport S.	94%	74%	14%	74%	3%	0%	0%	2%	0%
Telecom	100%	14%	14%	7%	14%	7%	0%	0%	7%
Banking/ Insurance	95%	17%	10%	4%	3%	1%	0%	0%	3%
Software	97%	20%	4%	20%	13%	10%	3%	3%	9%
Engineering Services	96%	44%	28%	16%	27%	9%	9%	24%	8%
Other Business Services	93%	42%	22%	23%	13%	2%	3%	10%	4%

Source: MIP-S / CIS II – Germany

The dominating role of IT becomes even more obvious when look at the share of companies, which do not name IT as important to innovation but only other technologies. Again retail trade stands somewhat out. However, even in retail trade only 7% of innovating companies innovate by using non-IT technologies another 6% retailers innovate without adopting new technologies at all. The majority of 57% of retailers use IT and any other technology. Similar observations can be made with the help of Table 3 for the other sectors. In the most IT prone sector like the financial services, Software and Telecom innovation without the use of IT seems to be nearly impossible.[12]

Table 3. Multiple Technologies in Use for Innovation in Service Sector

As share of all innovating companies in a given sector	**IT only**	**Only Other Technologies**	**IT and other Tech.**	**No New Technologies**
Wholesale Trade	13%	4%	75%	8%
Retail Trade	31%	7%	57%	6%
Transport	5%	4%	89%	2%
Telecom	86%	0%	14%	0%
Banking/Insurance	70%	0%	25%	5%
Software	43%	0%	54%	3%
Engineering Services	28%	2%	68%	2%
Other Business Services	43%	4%	49%	3%

Source: MIP-S / CIS II – Germany

Service sector innovations and the adoption and use of new technologies in services takes place although R&D seems to play a much smaller role in services than in the manufacturing. The German R&D statistics show that well below 10% of business sectors R&D is spend by service sector firms. The question then arise how can service sector firms maintain an equally large share of innovation companies. An obvious answer is that innovation in service occurs mainly in the form of embodied technology stemming out of manufacturing. However, this is certainly not the whole story. More in depth analyses show that the role of human capital in the development of innovation is considerably more important in service. This can be highlighted looking the skill structure of innovating and non-innovating companies.[13]

Table 4 shows the share of highly skilled, medium skilled and low skilled employees by sector. In all sectors the share of highly skilled employees in innovating firms is larger than in non-innovating companies. Moreover, in most sectors also the share of the medium skilled is lower in non-innovating companies than in the innovating ones.

Table 4. Skill Structure in Innovative and Non-innovative Firms

	Non Innovative Firms			Innovative Firms		
Skill Level	**High**	**Medium**	**Low**	**High**	**Medium**	**Low**
Wholesale Trade	8%	65%	28%	12%	71%	17%
Retail Trade	5%	76%	19%	12%	72%	17%
Transport	3%	63%	36%	5%	58%	37%
Telecom	n/a	n/a	n/a	32%	67%	1%
Banking/Insurance	15%	75%	10%	18%	70%	13%
Software	32%	57%	11%	48%	49%	4%
Engineering Services	43%	46%	11%	47%	46%	8%
Other Business Services	13%	61%	26%	26%	54%	21%

High = University or Technical University Degree or Masters Degree; Medium = Vocational Training; Low = No formal education (except school)
Source: MIP-S / CIS II – Germany

Moreover, Table 4 also shows that the share of highly qualified employees is largest in those sectors, which are more innovative and use more intensely use IT for the innovation. And this seems to be even more true for the innovating companies. This again points to the crucial role a sufficient human capital base plays for innovation in services.

In addition, the share of companies, which spend money on further vocational education and vocational training of their employee, is considerable higher in the sector with a larger share of innovating firms (see last column of Table 1). And firms investment in the expansion of the human capital base is also more important the more important IT technologies are for firms. Moreover, further education and training is also important for firms who do not innovate but invest in IT. So, it seems only possible to fully realize the benefits from investments in IT when complementary investments in human capital is made.

The crucial role of human capital for the adoption of new information technologies in services is also evident when looking at the share of expenditure on training in total innovation expenditure. This share exceeds considerably the comparable share of spending on training in manufacturing and amount for nearly the same share R&D has in manufacturing.

However, money spend by firms on the enlargement of the knowledge of their employees is not equally distributed across all employees. Employees with higher secondary and tertiary education levels receive more intense training opportunity than lower qualified workers (see Licht and Fier 1999). And this can be viewed as additional evidence for the fact that the adoption of IT also needs an increase of the initial skills at the level of employees. In turn, as will be shown in the next section low qualified employees represent the group of employees with suffers in their employment prospect due to the introduction of information technology in services.

4. THE IMPACTS OF INNOVATION ON THE DEMAND FOR SKILLS IN SERVICES

4.1 Innovation, IT and Growth

Before dealing with question whether innovation and new technology enhance the skill requirement of the service sector we shortly will deal with the growth of service sector in Germany and the role innovation activity plays in enhancing service sector growth and growth prospects. Firms will only enlarge their staff when their growth prospects are positive. So, when innovation enhances growth this will result in additional employment. In relation to this growth effect the efficiency enhancing aspects of innovation on the level and the skill structure of employment once the growth effect is accounted will be often of second order or will be only visible if the growth effect is eliminated. Therefore, a separate treatment of both effects is useful to understand this impact of innovation and technology on employment.

Overall, the service sector and especially technology-intensive and human capital intensive service sectors show above average growth in value added and employment growth rates in the last decade in Germany. This is also present in the innovation surveys were the majority of firms in services have increased their workforce and expect to proceed in this way in the future. So, the growth trend in service will prevail. However, when comparing the expansion of the German service sector to the growth of services in other countries, we recognize that Germany faces a less rapid expansion in service than most other countries. The main reason for this is the sluggish growth in the overall economy, which restricts also the growth prospects in services.

The growth also differs between East- and West-Germany and between firms of different sizes. West-German service firms grow much faster than East-German service firms. And the frequency of decreases falls with rising size class. In addition, there are considerable differences between the various branches. Banks and insurance companies in both regions had a high sales growth. In the W-Germany, transportation and telecommunications performed roughly as well as banks and insurance companies. Wholesalers and other business services led the mid-field, with retailing a long way behind. In the East-Germany, it is the software firms and the other service providers which follow the banks in the assessment rankings. Bringing up the rear in this part of the country are the wholesalers, and then, again a long way behind, the retailers. Another interesting fact is that those firms operating abroad (exporters, operating through partners or subsidiaries abroad) have had a significantly better sales trend. This is even more true for the innovating companies. By contrast, companies, which expect foreign competitors on the domestic market and are already encountering them, reported poorer sales trends than firms without any international competitors. As innovation is enhancing the export probability this is another channel for innovation to enhance the growth of a company. Of particular interest is the question of whether the companies using the various technologies covered by the survey stand out from the rest of the companies in regard to their growth.

In the multivariate analysis, it can be shown that, irrespective of the degree of internationalization, the size of the company, the particular branch of the economy involved, and the past sales trend, companies which have recently been using high-performance IT grow faster than their counterparts without intensive IT-use. The use of IT technologies appears in wide areas of the service sector to be a prevalent precondition for a positive sales trend.

4.2 Expected Changes in the Skill Structure and the Role of Innovation and IT

In order to demonstrate the impact of innovation and IT on the demand of skills we use instead of the actual demand for skill the expected change in the skill structure. The main reason for this is that a highly qualified workforce is a main requirement for innovation activities. Therefore, to address the problem of reverse causality we look at the expected change when linking employment demand to past innovation activities and technology use. Moreover, employment expectations can be meaningfully interpreted only against the background of the existing staff qualifications structure, which has already been addressed in section 3. Here we use a more refined qualification classification. We distinguish between college/university graduates with degrees in scientific/technical subjects, college/university graduates with degrees in economics, sociology or the humanities, professional school graduates ('technicians' and Germany's 'master degree' based on the vocational apprenticeship system), employees with a vocational qualification, and other employees.

An initial overview of general employment expectations in the various staff qualification categories is provided in the Figure 2.[14] For Germany as a whole, it can be seen that with the exception of "other employees" all qualification categories exhibit an upward trend. In each of the qualification categories, the majority of employees work for firms who wish to expand the category concerned. In the numerically most important category of employees, those with a vocational qualification (see the preceding section), the proportion of employees affected by a downward-trend expectation is greater than in the higher qualification categories. Nonetheless, the proportion of those who work in companies with rising employment prospects is considerably larger.

Prospects are gloomy, by contrast, for the "other employees", for a considerable number workers who are employed in companies with intend to downsize their payrolls. Conversely, roughly 25% of "other employees" work for companies, which anticipate increased hiring in this qualification category.

This indicates that there it is highly likely that a proportion of unqualified jobs has been lost in the service sector surveyed. And these are precisely those employees being already affected by above-average unemployment; indeed, the hope is frequently expressed that an expanding service industry may provide compensation here, but it is obvious that few stimuli can be expected from the sector of more company-oriented service providers.

More in depth analyses based on regression analyses can provide some additional insights which regard to the impacts of innovation and IT use. First of all, different trends are expected in the various size classes in the east and west of Germany. The three higher qualification categories tend to exhibit improved prospects with rising size class in the west, whereas in the east of the country, for example, the picture is precisely the reverse for graduates with a scientific/technical degree. No general trend emerges for the professional school graduate grouping. The prospects for employees with a vocational school qualification are best in both parts of Germany in firms with a workforce of 20 - 49, whereas firms with 50 - 249 on the payroll expect a lesser rise in the number of staff with a vocational school qualification. The negative expectations in the west of the country for employees with the poorest qualifications result from the employment plans in large companies.

When the employment prospects are considered separately for each branch, it emerges that positive prospects are confined solely to the software industry and other business services. Scientists have equally good expectations in all branches of activity, except in retailing, transportation and telecommunications. The economists and professional school graduates have equally good prospects in all branches, except for financial and engineering services. Employees with a vocational qualification exhibit lower net figures, though these are still positive except in the fields of engineering services. For the "other employees" grouping, the anticipated payroll changes are particularly extreme in banking and insurance, and in wholesaling/retailing. Otherwise, there are positive net figures only for the software and IT services industry and the business services, i.e. only in these fields can expanded employment for unqualified staff be anticipated.

In a further step, besides the distinguishing features already considered (branch, region, and size class), additional influencing factors are jointly examined on the individual level, like sales expectations, activities abroad, foreign competition, innovation status, use of new technologies, and staff qualifications structure. The aim here is to analyze whether company-specific characteristics can explain the increase in employees in the individual employee categories involved.[15] To do this we set up labor demand equations for each skill-group explaining the expected change in labor demand by sales expectations, relative factor prices as well as innovation activities and technology use. The results can be summarized as follows:

The sales expectations play a dominant role for employment increases in all qualification levels. Firms, which expect rising sales are more likely to hire staff from all qualification categories, as compared with otherwise equivalent firms who are not expecting their sales to rise during the years ahead.

Firms, which have introduced at least one innovation during the past three years plan to hire qualified personnel more frequently. Larger companies intend to increase their proportion of qualified personnel more considerably than do medium-sized and small companies.

Additional hiring of college graduates becomes more likely the higher the proportion of qualified personnel already is. By contrast, the hiring probabilities for staff with a vocational qualification and for poorly qualified applicants, drop the higher the existing proportion of qualified personnel already is.

The east-west differential is based less on different corporate structures and behavior patterns, but more on differing compositions of the branches concerned in terms of company size, innovation status and staff qualifications structure.

The use of new technologies per se exerts no direct influence on employment trends. This does however, also mean that new technologies can be expected to produce only slight productivity-related effects. On the contrary: these productivity effects can be implemented only in conjunction with the use of qualified employees. Firms with a more qualified staff more probably use IT technologies, which in turn increase the relative demand for skilled employees. The sales growth manifested due to the use of new technologies will, however, "trickle through" to employment trends as well.

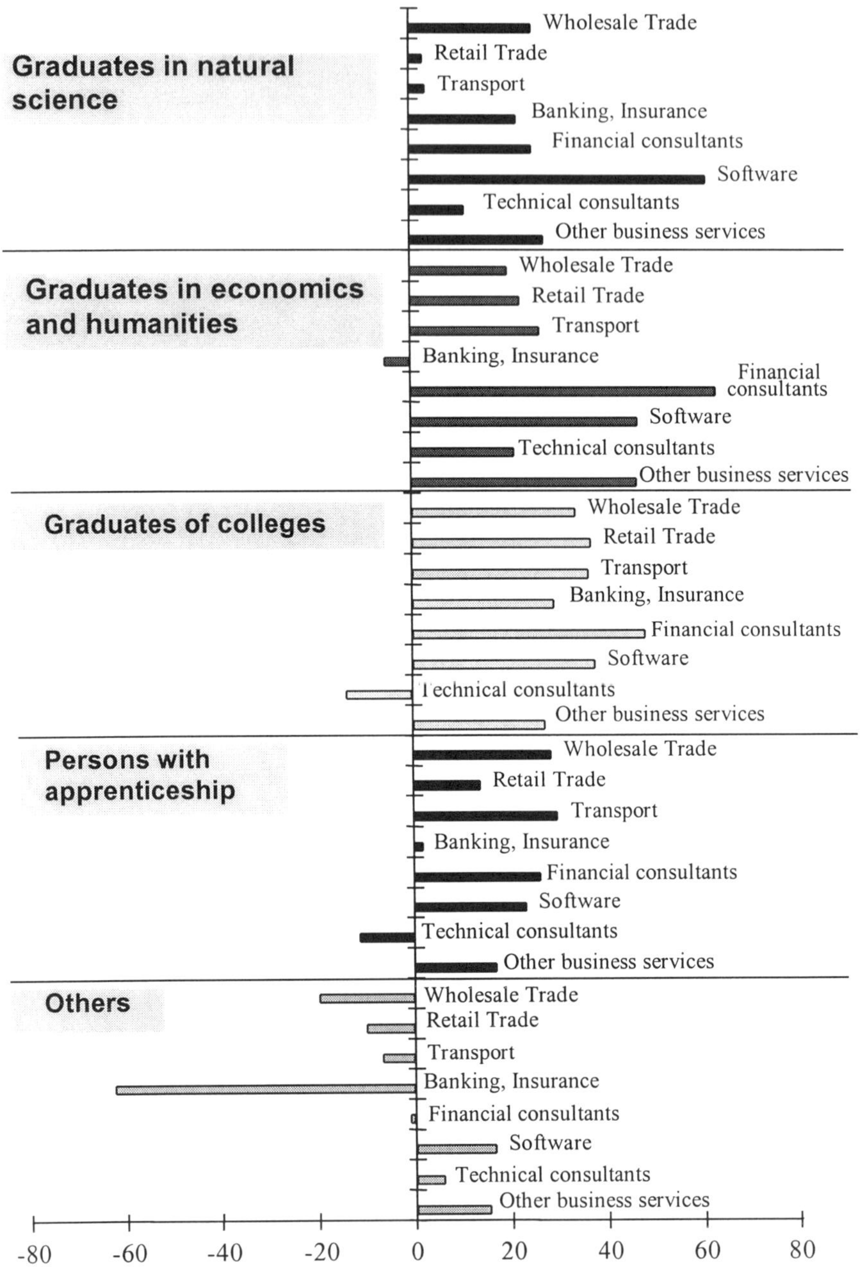

Figure 2. Distribution of Netted-out Employment Expectations by Industry and Skill Groups*

*Scale represents netted-out indicator of expected employment change.
Source: Mannheim Innovation Panel in Service; First Wave 1996; For explanation, see text.

The results indicate a sizeable shift in staff qualification structure within the service sector. The more highly qualified staff an enterprise possesses, the less likely it is to hire unqualified applicants. It cannot, however, be concluded that firms with higher employment expectations for highly qualified personnel simultaneously have declining employment prospects for unqualified applicants. However, we observe a negative correlation between the employment expectations for highly and poorly qualified staff. It is nonetheless advisable to devote greater attention to these shift trends, since if they intensify there will be far-reaching consequences for the labor market.

This is even more true as the present trend in provision of continuing training and continuing education by companies to their employees is strictly biased towards highly qualified and already skilled employees. The innovation process is additionally less markedly de-coupled from other corporate performance processes than is the case in the manufacturing sector. Attention should accordingly not be confined to those training expenditures directly linked to innovation projects. The high degree of interactivity in the process of goods/service creation means that qualification levels and training become closely related to innovative activity and technological/organizational change in the service sector as such. However, innovating companies invest significantly more in staff training than do non-innovating companies. This is also true for companies with a high proportion of highly qualified personnel spend more per capita on training schemes. This means that a high level of qualification is a precondition for further training / up-skilling measures in the companies.

The use of new technologies and also the proportion of ICT-related investments in the overall investment budget exert a markedly positive effect on the level of training expenditures. Therefore, it seems dangerous when mainly the already-qualified staff profits from training (due to technological change). Since this implies that the unqualified employees will thus be permanently excluded from the labor market, and the employment chances for the poorly qualified, slight enough as they are, will deteriorate still further as technological change progresses. Reforms of the educational systems are called for especially as a good primary education is the ticket on which further vocational education and training is based.

5. SOME CONSEQUENCES FOR INNOVATION POLICY

The employment expectations in the German service sector can be rated as positive overall, at least for employees with vocational training or university degree. In companies with innovative activities, there will be more jobs generated for qualified staff. There are also indications that the staff qualification structure is going to shift towards favoring more highly qualified people. This highlights the role of human capital in the productive process of the service sector segments examined, and underlines the necessity for helping the less well qualified by implementing training schemes and/or changes in the education structure.

Corporate training schemes are aimed at keeping abreast of technological change. Companies with a higher proportion of highly qualified staff and users of new technologies exhibit more intensive efforts in terms of training. There are almost no discernible indications that companies are specifically attempting to familiarize poorly qualified staff with the utilization of new technologies. It is more probable that the requisite technical know-how is being acquired through hiring new employees, and that in-house training schemes are primarily benefiting qualified and highly qualified staff. The differences in human capital among employees are thus being reinforced by training approaches during the course of ongoing technological change.

Information and communication technologies constitute the pre-eminent cross-section technology for the service sector, ahead of all other modern-day technologies. Small and medium-sized enterprises of the service sector's traditional segments (e.g. carriers and forwarding agents and trade) often do not possess the internal resources to exploit the opportunities these technologies offer. On the one hand, technological acceleration also causes any accumulated know-how to become rapidly obsolete, while on the other hand increasing liberalization demands intensified use of precisely these technologies in order to maintain competitiveness. This, however, reinforces the difficulties SMEs face in coping with the diffusion of IT in traditional service sectors. In addition time pressures on the side of the key personal in SMEs in the booming business services prevent them from reinvesting in human capital in order to keep path with the path of technology. Innovation policies should accordingly aim at increasing these companies' absorptive capabilities.

This can be exemplified by the transportation segment: following complete liberalization of the transportation segment throughout the European Union in 1998, the networking of local, national and international carriers, plus more intensive use of I&C technologies, will attain central importance for safeguarding competitive positions. The decline in prices for freight services, already triggered by initial steps towards liberalization, is hitting the liquidity of medium-size carriers, and is often preventing systematic efforts towards utilizing information and communication technologies and the requisite restructuring programs.

The know-how available in public research institutions is important for many small and medium-sized service enterprises planning to use new technologies. Formal arrangements for cooperation between SMEs and research institutions, however, are relatively widespread in only a few segments. Since innovation research has found that cooperative arrangements comprise essential and frequently effective sources for know-how transfer, we must ask how more such cooperative arrangements can be stimulated. The primary aim here is by means of altered priorities to develop appropriate options at the universities and research institutes, who in their R&D activities are still very largely focused on industrial production/development processes. Consideration must also be given as to how the often inherently intangible character of service products can best be allowed for, since precisely for this reason existing programs tend to pass them by.

Many service companies possess excellent preconditions for cooperative arrangements with research institutions. Software companies and consulting engineers, particularly, thanks to the high qualifications of their workforce, are often unaffected by the most important barrier to cooperation with bodies from the public-sector infrastructure: unfamiliarity with the idiosyncrasies and customs of the university and academic communities. Formal and informal arrangements for cooperation are thus encountered quite often in these branches. Service companies of this kind are also excellently suited for acting as a driving force to expand further relationships between public-sector research institutions and small or medium-size enterprises from the manufacturing sector. Research and technology policies should, when formulating promotional measures for SMEs in the manufacturing sector, attempt to exploit this function of service-sector companies, and take more account of service companies' technology transfer function. In conjunction with further diffusion of information and communication technologies, particularly, new service segments like information mediators will emerge.

Our results indicate that in other branches of the service sector as well, especially when small and medium-sized enterprises are involved, there are substantial problems to overcome in adapting modern I&C technologies. The start-up costs for these new technologies include not only the capital investment directly required for this purpose, but also the substantial indirect outlay for generating the requisite human capital.

In view of the severe problems of unemployment in the Federal Republic of Germany, the above-mentioned development in staff qualification structure constitutes a central field of action. The transformation into a service society will entail a substantial expansion of jobs for qualified and highly qualified employees. This growth of "good jobs" will by far exceed the increase in "bad jobs", if indeed there is any increase in these at all in the modern service. The change in qualification structure towards more highly qualified employees, well documented in the manufacturing sector over recent years, is being manifested as a similar process in the service segments surveyed. Political measures for combating unemployment have to take account of this trend, to ensure that unemployment and underemployment do not become increasingly a permanent condition for unqualified workers; not least because long-term unemployment and technical progress will continue to devalue existing qualifications.

Efforts to improve employment chances for unqualified workers can address both the relative prices for unqualified work and the options for increasing these workers' productivity. Various policies can be applied separately or in combination To decrease the relative prices for unqualified work,: option include (1) reducing the non-wage labor costs to be paid by the companies for these workers, e.g. by cutting mandatory social security contributions, or (2) reducing earned income for unqualified work, a step whose effects on available income can be cushioned by tax exemptions.

Training schemes could enhance the productivity of the low qualified employees At present training schemes tend to benefit mainly the qualified employees. Expansion of corporate training programs, however, simultaneously demands improvements in the quality standards of the schemes involved, so that the search for suitable bodies to sponsor training schemes is facilitated both for companies (and hcrc thc SMEs in particular) and for employees. However, it is hart to say whether a tighter regulation of corporate training efforts may be beneficial. In view of the multi-faceted, fast-changing context entailed by new technologies, direct regulation of training schemes would appear of limited assistance, involving as it does the risk of restricting the requisite flexibility in the training options on offer.

In any case an increase the educational attainment of employees is a key for modern economies to profit from the job engine represented by the knowledge base service economy. Moreover, educational reform should be seen as an integral part of innovation policy and should not be treated separately. As long ago recognized by economists (see e.g. Nelson and Phelps 1966, Aghion and Howitt 1998) see at the macro-level and at the micro-level of firms a well-educated workforce is an important driver of productivity and employment growth. At the individual level of employees and unemployment investing in education and life-long learning presents a first best strategy to drive down the expected individual unemployment probabilities. Given this potential utilities involved for all parties we should start with changing the priorities. Lower the relative cost of education and vocational training (see Dohmen 1999 or OECD 1998 for comparisons of the rates of return on physical and human capital investment) will certainly increase the investment in human capital financed by the individuals, the firms and the government.

Notes

[1] I would like to thank Thorsten Doher, Günther Ebling, Norbert Janz and Hildrun Niggeman for their ongoing efforts in preparing the innovation data set used in this study. I also thank Engelbert Beyer at the German Ministry of Education and Research for stimulating me to work on the relation of innovation in services and their labour market impacts. Financial support by the DFG is gratefully acknowledged.

[2] However, international comparisons of employment shares are sometime misleading when hours worked are not taken into account and some employee groups are systematically not included in this comparison. This is relevant with regard to the so called 630-DM-jobs in Germany (see DIW 1998 for more details and some alternative calculations of German US comparisons).

[3] See for details on the measurement problem a recent special issue of the Canadian Journal of Economics (1999, Vol. 32 April).

[4] The MIP-S was commissioned by the federal ministry of science and technology (*BMBF)*. The *Fraunhofer Institut für System- und Innovationsforschung (ISI)* and *infas* a private company specialized in conducting surveys are also involved in this project. The concept of the MIP-S and further results are presented in Licht et al. (1996).

[5] This seems to be the best available data source. No official register is available for service sector enterprises. The most recent data from the statistical on e.g. firm size or industry distribution of service sector firms is for 1987. The Federal employment office publishes data based on the establishment level. However, this data only covers establishments with at least one employee, which is subjected to mandatory membership in the employment security system. This data only comprise figure on the number of establishment and the number of employees with mandatory contributions to the social security system. No sales or other data are available. Also, no data on firm's addresses etc. is available to third parties.

[6] As telecom deregulation in Germany leads to a strongly rising number of telecom service providers in the second half of the nineties additional telecom companies were integrated in the second wave making a separate treatment of the sector possible.

[7] Although the coverage of sectors is somewhat different Baldwin et al. (1998) report lower figures for the share of innovative firms in Canada. Moreover, in most European countries (except for Sweden and UK) the share of innovative firms is also lower than in Germany.

[8] We will discuss these issues only shortly and refer the reader e.g. to our recent book (Licht et al. 1997) or the paper by Ian Miles in this volume.

[9] Innovating companies are defined – according to the OSLO manual – as companies which introduce new or significantly improved products and implement significantly new production process within a three years period (1994-1996).

[10] Data for these aggregates refer to the year 1997. For details see Janz et al. 1999a and 1999b. Detailed data for 1997 are not yet available.

[11] The list is more detailed than the technologies depicted in Table 2. In the case of information technologies it considers of the usage of mainframes, workstations, PCs, packaged standard software, data bases, high quality net works (e.g. ISDN) or multimedia technology. Environmental technologies comprise e.g. recycling, energy saving techniques, pollution control measurement.

[12] However, the observation that IT and other technologies often occurs in the same firm at the same time not necessarily is an indicator of technology fusion going on. It can well be

the case that two separate innovations are made one using IT and the other using e.g. environmental technologies.

[13] The reverse causality – innovation influences the skill structure – is inferred in the next section in more detail.

[14] This figure is constructed in the following way. We first calculated the share of employees in each skill group who is working for companies, which expect to expand their workforce in this skill group. Then we calculated the skill group specific share of employees working for companies, which plan to contract the workforce in this skill group. Finally we net out the employment expectation in each skill group by subtracting the number of employees in contracting firms from the number of employees in expanding firms. The resulting number is a good proxy for the expected growth rates as evident from the business cycle research has proven (Carlson and Parkin 1975 and Batchelor and. Orr 1988). Therefore, the larger this number is the higher is the expected employment growth.

[15] This is done using separate order probit regression models for each skill group. Kaiser (1998) uses a more refined econometric method considering the correlation between the error terms of these equations leading to more efficient estimates. His results confirm the interpretation made here.

PART IV

IMPACTS

Chapter 9

The Impact of the Information Revolution on the Global Corporation

CLIFF WYMBS
City University of New York

1. INTRODUCTION

Much has been written about how the Internet removes time and locational constraints associated with global commerce (Wymbs, 1997). Traditional studies describing international activity focus on how firms and individuals from two or more cultures, countries or regions are brought together to accomplish a particular business purpose. The Internet, as an informational exchange medium, is beginning to be used in international business; however, the specific areas where it creates value are just being defined. The intent of this study is to identify those areas where the Internet via electronic commerce will have the greatest effect in changing the relative components of the value chain and then explore areas of international business (International Trade, Foreign Direct Investment and Strategic Alliances) which likely would be early users of the Internet. Research questions that will be partially addressed in this analysis include:

- Does the current information-intensive environment represent a fundamental change in the business paradigm of just five years ago?
- If so, how will electronic commerce transform business and redefine the nature of products and the form of deliverables?
- How will electronic commerce affect trade, foreign direct investment and strategic alliances of global firms?

Before these questions can be analyzed, the domain of electronic commerce on the Internet must be defined. More specifically, answers to the questions of what is it, who is using it and why, and what are the problems of electronic commerce and what must be done to solve them, provides the necessary context for the analysis.

2. ELECTRONIC COMMERCE

Over the last decade, a number of events that are driven by a new generation of telecommunication advances suggest that the world economy may be entering a new phase of market-based capitalism (Dunning, 1995). Today, thousands of businesses throughout the world have been combining the functional capabilities of the computer and telecommunications to engage in electronic commerce, i.e. the use of the Internet for commercial transactions. Because companies now exchange information and products electronically rather than sending and receiving paper documents, they are experiencing remarkable and unparalleled improvements in accuracy, speed and efficiency with which commercial transactions may be negotiated, confirmed and performed (Ritter, 1992). Ritter further states that when electronic commerce is used for international transactions, it is able to overcome many of the traditional barriers associated with international trade presented by geographic, lingual and cultural disparities between possible trading partners. In addition to acting as a market facilitator, electronic markets are changing the form of deliverables and characteristics of products (i.e. those which can be 'housed' in telecommunication related hardware).

McKenzie and Lee (1991) stated that the current revolution in computer and information technology has enabled firms to move many assets around the world at the touch of a button. (However some intangible assets such as knowledge remain locational sticky.) It has caused an explosion of competition between businesses and has forced governments to reduce spending, to decrease regulations and to lower taxes in order attract business investment. Information and knowledge have been merged to become today's society's new capital. These authors further assert that information, now no more physical than electrical blips on the computer or disc, remains fluid and difficult to contain in any one place and stands ready to leap across continents via satellite or fiber optic cables. Knowledge, embedded in human capital, is rapidly replacing physical items, e.g., steel, as the most important source of economic power. New technology reduces space and time constraints and is increasing the ease with which information flows among people around the world (Wymbs, 1996). However to take advantage of these trends, each country requires a critical level of spatially sticky infrastructure (immobile assets) to serve as the glue necessary to connect the mobile information and knowledge assets. Clearly, markets are becoming global and organizational structures and business processes based on the old paradigm have to change.

According to Maloff (1995), the Internet is changing commerce as surely as did the telephone or railroads. Now for the first time, employees located in most parts of the world can communicate with one another in a timely and cost-effective manner. Customers in Chicago, Paris and Hong Kong can view products and buy them without leaving their home, while sales people can access their computer system, conduct a demonstration hundreds of miles away, and instantaneously enter orders. Adam et al. (1996) pointed out that the world has become a smaller place, with telecommunications technology making terrestrial distances insignificant and transportation technologies shrinking physical distances. The transformation of the world into a global village has caused, and will further cause, revolutionary changes in physical and social infrastructure rivaling those of the industrial revolution. According to Newing (1997), history tells us that companies which dominate before a fundamental paradigm shift are not likely to do so after the shift. In the same article, Mr. Steele of IDC states that different approaches are needed on the Internet and the winners will likely be those who have no legacy.

2.1 What is it?

According to Maloff (1997) the basic communications needs of virtual or real businesses have not changed, *i.e.*, they need to communicate with each other internally (employee relations, distribution of company news, etc.), with customers (market information gathering, competitive assessment, image creation, closing orders, etc.) and with suppliers (bills of lading, accounts receivable, etc.). Historically, businesses have dealt with these needs via face-to-face meetings, mail, telephone, fax and courier services, while today, the Internet has opened up many new cost-effective electronic alternatives. Similarly, Ritter (1992) states that with electronic commerce, the general nature of the information required does not significantly change, nor do the needs of the participants for the information; however, what occurs is a transformation of the means employed to create, store and process the information, and the elimination of duplicative efforts in those activities. A new class of commercial and administrative players are created. Electronic networks used to carry electronic data interchange (EDI), facsimiles, electronic mail, etc. have been constructed to facilitate communications between the new stakeholders.

Adams and Yesha (1997) note that electronic commerce supports effective and efficient business interactions and transactions that take place on behalf of consumers, intermediaries, and producers. Along the same lines, Ritter (1992) portrays electronic commerce as a disciplined way of doing business because of the inherent needs of all parties to create, transmit, process and store information electronically in a rational manner. Because of an industry's ability to set standards to facilitate use of technologies in electronic commerce, most electronic commerce will likely evolve on an industry by industry value chain basis, *e.g.*, banks, automotive manufacturers, transportation companies, etc. More specifically, Adams (1994) includes under the broad definition of electronic commerce a number of groupware programs, such a Lotus Notes and WordPerfect Office, as well as public E-mail networks, such as MCI mail, CompuServe, and the Internet.

Maloff (1997) describes the Internet as an empowering tool that enables organizations to grow without relocation expense, to advertise and promote themselves less expensively to a global audience, and to create virtual companies where workers are independent of location. Bay (1995) asserts that the Internet is the death of distance and in the near future may also eliminate usage time, connectivity and time and distance. Today businesses spend billions of dollars on maintaining their worldwide proprietary internal networks based on leased lines. Someday, the Internet can provide the common infrastructure platform and replace this. Kelly (1997) expands on this concept and concludes that the Internet is a prototype of the global information infrastructure that will lay the platform for the electronic commerce of the 21th century and serve as the fundamental value added component of the global corporation. The significance of the Internet is not so much what it is but what it will become.

2.2 Who is using it and why?

There is no universal agreement on Internet usage. Shafe (1997) found that the Internet was used by 40 million people in 1996, while Margherio (1998) estimates that more than 100 million people were using Internet by the end of 1997. As of December 1996, there were about 630,000 Internet domain names registered. By the end of 1997, the number of domain names increased to over 1.5 million. Today more than 55% of large US companies have web sites.

Blundon (1996) cites the Georgia Institute of Technology's Graphic, Visual, and Usability Center (GVU) which reported detailed demographics associated with Internet Users. The GVU April 1996 survey found that United States respondents represented 73.4% of the total, Europeans 10.8%,

Canadian and Mexican 8.4% and Asia and the rest of the world 7.4%. Compared to the same survey a year ago where the United States accounted for 80.3%, this represents a shift away from U.S. dominance in total Web usage. Greston (1996) found that there is no single Internet development path in Europe. Each country is tailoring the service to its own resources and culture. It is hoped that as telecommunications costs come down in Europe, EU countries will embrace the Internet, an engine and symbol of globalization, and not be left at a competitive disadvantage vis-a-vis the United States and the leading Asian economies. Mesher (1996) cites that Asian Internet usage lags behind Europe and the United States, but it is developing rapidly. In 1997, there were more than three million users in the region, and the total is growing at about 5% per month. Japan, with 1.5 million users, is by far the largest user. Because of the high local telephone charges in Japan ($.90 per minute) many users wait until midnight to log-on when the telephone rates are the lowest. To service customers more cost effectively, many cable TV operations in Japan are now offering flat rate Internet access. Some governments, such as China and Vietnam, attempt to control Internet access. In other countries, *e.g.*, Philippines, Thailand and Indonesia, language barriers, low living standards, poor telephone infrastructure and lack of network technicians will likely limit Internet development. Blunden (1996) found that the average age of the Internet users was 33 years and women now account for 31.5% usage, up from 29.5% last year. The average income of a user is $59,000 per year, but this is lower than $69,000 a year ago, probably attributable to the increased student usage of the Internet. In descending order of occupational usage, education accounts for 30% of the usage, computer-related jobs, 27.8%, professional, 18.9% and managerial, 10.7%. Bottom line, the Internet audience is well educated, well-healed, and younger than the national average.

Bransten (1997) found that the Internet retailers vary from thousands of small Internet only companies to the world's largest retailer, Wal-Mart. Both in the UK and in major cities in the US, supermarkets have been going online. However, the most popular items purchased via the Internet are: Computer software and hardware, travel related items such as airline tickets, and leisure goods such as CDS, books and videocassettes. In addition to retail, Yoder (1995) cites a Forrester Research study that predicts the number of online security trading accounts going from approximately 400,000 in 1994 to more than 1.2 million in 1998. McCann-Erickson found 1996 advertising revenues on the Internet to be between $150-200 million, a small percentage of the United States advertising expenditure of more than $166 billion; however, rapid growth is forecasted with a year 2000 Web advertising estimate of $2 billion.

In Blunden (1996), both a Computer Intelligence Infocorp and a Nielsen Media Research survey found that approximately 10% of Internet users, 2.6 million, shop or obtain commercial services such as banking or travel on the Internet. ActivMedia and Forrester Research are in reasonable agreement that current sales on the Internet amount to approximately $500 million while their forecasts diverge with ActivMedia expecting $46 billion in sales in 1998 and Forrester only $6 billion. Bransten found IDC's forecast to be considerably more bullish at $95 billion. Irrespective of which forecast is accepted, the Internet is a rapidly growing portion of the US retail market estimated by the National Retail Federation in 1996 to be $2,400 billion. There appears to be great potential: Krumenaker (1996) states that although there were about 150 million PCS in use at that time, only about six million are connected to the Internet. Bay (1995) boldly states that in the not too distant future every personal computer will be connected to the Internet on a worldwide basis.

Answering the why question is easier: Efficiency, cost reduction and providing better overall customer service are the clear results of using Internet. Bransten (1997) found the main attraction of shopping over the Internet to be its efficiency, particularly computer software and books where you can download and view the item before purchase. Ubios (1997) states that its greatest potential is in certain transactional services, *e.g.*, airlines, hotel and rental cars and, soon, banking. Other transactional services that are complex and require extreme security will stay on proprietary EDI systems. Similarly, Adam et al. (1996) state that the technology of globalization is being driven by commercial incentives for improving efficiency of business enterprises as well as societal concerns with improving quality of life. Adam et al. quantified this, stating that savings to the federal government of shifting to a paperless system for recording, managing, and maintaining business records is estimated at $200 billion annually. According to Mr. Einnsman, head of P&G in Europe, because the Internet permits companies to communicate directly with customers, it will likely cut distribution costs, provide more detailed information about individuals' needs and purchasing patterns and allow manufacturers to strengthen brand loyalty (De Jonquieres, 1997). However, to gain these benefits, companies will have to radically change their traditional approach to the mass market and develop techniques to better listen to what consumers say and become their problem solvers.

2.3 What are the problems of electronic commerce and what must be done to solve them?

Adam and Yesha (1997) describe the problems raised in building a global information infrastructure as a mix of technology, public policy and culture.

Nairn (1997) stated that tremendous growth is putting severe strains on the Internet's infrastructure. More specifically, Internet traffic has been doubling every 100 days (Scace, 1998) which reflects both a widening and deepening of information transmitted. To help solve this problem, many Internet service providers are moving toward advanced networking technologies such as frame relay and asynchronous transfer mode to remove their dependence on routers. Adam et al. (1996) point out the shortcoming of the Internet in its current state: lack of security, although secure protocols are being considered; minimal or nonexistent quality of service guarantees; no mechanisms to protect intellectual property; and no support of interoperability or data interchange standards.

Achieving solutions will require the cooperation of many disciplines, *i.e.*, law, economics, business, education, sociology, psychology, entertainment, computer science, etc. (Adams and Yesha, 1997). Manchester (1997) points out that meteoric Internet growth can only be accomplished in a secure networking environment between the retailer and the purchaser, regardless of international borders. Newing (1997) believed that 1998 would bring necessary security technology to eliminate today's most pressing problem to the successful implementation of electronic commerce and that problems associated with international product liability and problems at the government level over taxes and import duties would also be mitigated.

Ritter (1992) identifies the ideal business environment for electronic commerce to include universal adoption of international standards, both as communication protocol and message development, and comparably low costs and universality. He believed that time and appropriate legal rule making would make this ideal environment a reality. Adam, et al. (1996) cited many reasons for the regulation of the Internet: to provide common standards and security for global users; to balance freedom of speech against offensive and libellous use; to provide protection (copyright) for knowledge providers; to protect countries, languages and cultures; to ensure access for both rich and poor individuals and countries; and to encourage educational and socially beneficial uses.

More specifically, the digital revolution and electronic commerce is changing the respective roles of government and the private sector (Margherio, 1998). For the last one hundred years, the role of government has been to build and regulate infrastructure. Governments granted monopolies in telecommunications and power, gave land for railroads, and subsidized airlines and the emerging Internet. The rapid pace of technology change and the borderless world being created by electronic commerce dictate a new paradigm for government and business interrelationship. Margherio (1998) identifies the optimal governmental conditions for the digital economy to grow: Governments must allow electronic commerce to grow up in an environment driven by markets, not burdened with excessive regulation, taxation and censorship. Where possible, rules for Internet and electronic commerce should result from private collective action, not government regulation. Governments must support the creation of a predictable legal environment globally for doing business on the Internet, but must exercise their control in a non-bureaucratic fashion. Greater competition in telecommunication and broadcast industries should be encouraged by government so that high bandwidth services are brought to homes and offices around the world and the new converged markets of broadcast, telephony and the Internet operate on laws of competition and consumer choice and not government regulation. Governments should not place discriminatory taxation on the Internet. The Internet should function as seamless global marketplace with no artificial barriers erected by governments.

As with other major social transformations, e.g. the Industrial revolution, much economic dislocation is likely to be created by the digital economy. Bottom line, the private sector and government must work together to minimize dislocations through progressive human resources policies, make the Internet a safe environment and prevent abusive invasions of privacy and not impede its commercial development. The capstone question of the analysis follows.

3. HOW IS ELECTRONIC COMMERCE TRANSFORMING BUSINESS?

Two separate inquiries are necessary to answer the capstone question of this paper. The first pertains to how the emerging information-rich environment is changing the basic business paradigm and the second deals with the systematic analysis of the changes due to the emergence of electronic commerce.

3.1 Does the 1998 information-intensive environment represent a fundamental change in the business paradigm of just five years ago?

History provides us guidance for the first inquiry. Two economists, Joseph Schumpeter and John Dunning, describe drivers of the last business paradigm shift that occurred at the turn of the 19th century. Schumpeter

(1934) asserted that innovation and technological advances are the major causes of economic growth and change in the business environment. Dunning (1988) indicated that during this period a new wave of technological advances occurred that in many ways was more profound and far reaching than its predecessors. These advances stem mostly from the United States, and very much reflect both its particular factor endowments and its market characteristics. They were stimulated by the creation of a new communication network, which helped shift both the demand and supply curves for goods to the right. "Such technological changes fundamentally affected the production of firms, their capacity for and strategy of growth, and the market environment in which they operated. They introduced new kinds of competitive advantage, which affected both the ability of firms to exploit foreign markets and the form by which advantages were exploited." (Dunning 1988, p88.) The parallels to today's environment are quite remarkable.

Unsal (1997) observed that over the last decade, information and communication technologies have further revolutionized global business by eliminating time and distance impediments. Since April 30, 1995 when the National Science Foundation opened the Internet to commercial enterprise, firms have been increasingly using the World Wide Web (WWW) to change business processes associated with marketing, supply chain management, public relations, customer support, product sales, and electronic data interchange. The Internet is rapidly becoming a key resource for international corporations as they position themselves for operating in the next Millennium. It is radically affecting the way business distributes information and communicates with foreign affiliates, domestic partners and business clients. The emergence of hybrid uses of the Internet -- Intranets (an internal network that uses Internet software and technology) and Extranets (a system that allows those outside an internal network to access information and data stored on the Intranet) are redesigning the business landscape and how value is delivered to customers (Shoa and Shoa, 1997). These networks have many desirable properties, namely, being relatively inexpensive to implement, secure, easy to use, and able to accommodate emerging applications, that encourage meteoric growth and the changing of the fundamental business organizational and operational paradigms.

Similarly, Evans and Wurster (1997) observed that a fundamental shift in the economics of information is happening -- a shift that will cause change in the structure of entire industries and the way in which companies compete. The shift is greatly facilitated by the widespread use of Internet Technology, but technology is only part of the story. More important are the millions of people who are using the technology and how their behaviors are changing the value chains for business in many sectors of the economy. This explosion of connectivity, driven by the Internet, separates the physical and informational parts of the value chain, and will surely result in a plethora of value chain combinations and permutations. Each time information is substituted for physical connectively, a business opportunity for an entrepreneur and a threat of survival for an entrenched firm will result.

In summary, the two key differences between business commerce today and five years ago is the reduction of transactional friction among business linkages due to an advanced communication medium and the amount and sources of information available at little or no cost. These two somewhat simple conceptual changes are powerful enough to drive fundamental change in organizational structure. The first, lower transaction costs, has the effect of reducing the economics associated with hierarchal organizational forms and encouraging the emergence of distributed networks. The second change, increased information, permits a flatter organization with less need for information gatekeepers. McKeever (1996) re-enforces these points by stating that Internet-related electronic commerce is the great enabler, linking all parts of the technology value chain and streamlining the configuration, ordering, and support of the thousands of products that must be integrated to build a viable customer solution. It is a great infrastructure tool making more efficient the supplier to distributor to reseller to end-user. Both of these changes work together to decrease the pyramid form and level of vertical integration that had become the accepted norm over the past 100 years. However, many of the very same informational factors that reduce the need for vertical integration encourage horizontal integration within industries and across countries; therefore, the effect of the new informational age paradigm of size requires further study and is indeterminate at this time.

4. INFORMATION AGE VALUE CHAIN

Because the Internet is both an infrastructure enabler of business and a separate channel of commerce, the analytic approach must be able to address how specific information links are substituted for physical links. One such approach that systematically evaluates the effect of an informational paradigm shift is the value chain analysis. Michael Porter's version is composed of nine components, five primary activities (inbound logistics, operations, outbound logistics, marketing and sales and service) and four support activities (firm infrastructure, human resource management, technology development and procurement). The information revolution and the digital age are dramatically changing the relative importance on each of these nine factors in most industries. Understanding and weighing the components and linkages of the new *information age value chain* is extremely important because it affects organizational and strategic decisions, which in turn directly affect a firm's profit margin. Pre-Internet, Japanese firms were the first to use inter-firm information links as substitutes for inventories in the just-in-time manufacturing process. The Internet has the potential to have a much greater effect by electronically linking customers and manufacturers and significantly altering outbound logistic and marketing functions of a firm; facilitating cooperative arrangements; removing inter-firm physical bottlenecks; redefining the competitive playing field. In attempting to demonstrate these effects, first the justification for an information value chain is explored. This is followed by a redefinition of the boundaries of the firm to fit information environment. Next the E-Commerce value chain is defined and contrasted with the Porter value chain, then new business processes are defined for the E-Commerce world. This is followed by specific E-Commerce applications. Finally an assessment of applicability of existing theories to information age is performed.

4.1 Justification for the Information Value Chain

Many of the components of the Porter value added chain have become separate self-contained networks, e.g. outsourcing a firm's data networking, and the critical decision associated with interconnecting these networks physically or electronically is rapidly becoming the key question senior managers must address. The Internet is rapidly becoming an important new channel for commerce because it allows for direct, ubiquitous links to anyone, anywhere. It lets firms build interactive relationships with customers and suppliers, and deliver new products and services at very low cost (Ghosh, 1998). Not surprisingly, start-up companies, rather than large businesses with established brands and physical distribution channels, have been much more likely to exploit the new technology. They have no existing value chain to protect and can tailor their business to the Internet. Ghosh (1998) indicates that these upstart entrepreneurs tend to leverage one of four possible electronic opportunities:

1. Direct linkage to whomever (buyers, suppliers, customers, etc.), thereby significantly reducing transaction costs;
2. Value chain bypass -- producers like Dell Computers can communicate directly to customers without distributors;
3. Development of new products and services, only limited by the creativity of entrepreneurs;
4. Access control and establishment of de facto standards for a particular industry or segment, a company that moves first is a threat to those who do not.

A customer-centric focus appears to be the common thread associated with above opportunities. To be successful in this photon-connected world, the customers must receive at least one of the following service attributes at a better level, more convenient, interactive, personalized or something new. Many electronic services displace functions associated with increasing marginal cost in the physical world such as sales, marketing and service, with declining and in some cases near-zero marginal cost in the cyberworld, e.g., an incremental customer visiting at an electronic storefront.

4.2 Firm Redefinition

The defining of the Internet as a channel somewhere between a hierarchy and the market coupled with the linkage of value chains among firms makes the definition of a firm boundaries and functions difficult and evolving concepts. Also, in the emerging knowledge-based economy, access to information is critical to accurate and real time decision making. Trading partners can no longer be viewed or treated as separate, distinct entities (Chizzo, 1998). They must possess tools to communicate and function in real time and in an integrated effort to serve the customer. Therefore, traditional criteria based on control and ownership are not particularly applicable in an era of knowledge capitalism when a firm's key resources, human capital, vote daily on where to be deployed and when its critical customer and supplier linkages are increasingly outside rather than within the firm. Maybe a better way to characterize the firm is as a networked activity set (NAS), guided by management to effectively target and compete in a particular product domain. Key dimension components of this activity set include:

- *Concept:* Focuses on leadership of the NAS. The leader must have a clear vision of the market domain in which the firm wishes to participate, what internally it brings to this process and what should be sourced from outside.
- *Creativity:* Centers on turning new ideas into good currency. NASs must continually reinvent themselves and possess entrepreneurial spirit of Schumpeterian "creative destruction." It requires marrying a firm's extended-network capabilities with an understanding of the customer and supplier requirements.
- *Capabilities:* Consists of both objective and tacit knowledge of the NAS employees and its information and social systems that tie them together with both internal and external players.
- *Cooperation:* Is multifaceted dealing with at a minimum inter-department, inter-business unit, and inter-firm, and firm-customer/supplier linkages. This dimension has gained in importance as traditional hierarchical

command and control management structures are being replaced by flatter and more distributed systems. In fact, some virtual companies "resellers" compete solely by coordinating the production process of others.

- *Coherence:* Relates to a unified sense of purpose embedded in shared values and assumptions. It has become the glue that keeps both the internal fabric of a NAS together and the web of alliances functioning with a consistent direction.
- *Connectivity:* Brings together a seamless array of activities focusing on a common purpose. It can be thought of as the great enabler. If used effectively, it facilitates creativity, capabilities, cooperation and coherence.
- *Capture:* Is what is required to win customers' loyalty, which can no longer be taken for granted. Customers must be sought out and offered superior value propositions. Each customer is a battle and each purchase is a capture.
- *Competition:* Is any other entities' NAS that impacts or could impact you or your customers value chain.

4.3 E-Commerce Value Chain

A networked activity set definition of a firm leads to much more interaction between consumers and producers throughout the value chain process. In fact, the E-Commerce Value Chain consists of both a producer and consumer branch with almost as many linkages as nodes on the individual branches. Its focus is on speed, ease of operation and customer satisfaction with information as the facilitating dynamic. IBM, one of the leaders in the electronic commerce movement, also views the value chain as a dual process. (See Figure 1.)

Figure 1

Value Chain Electronic Commerce

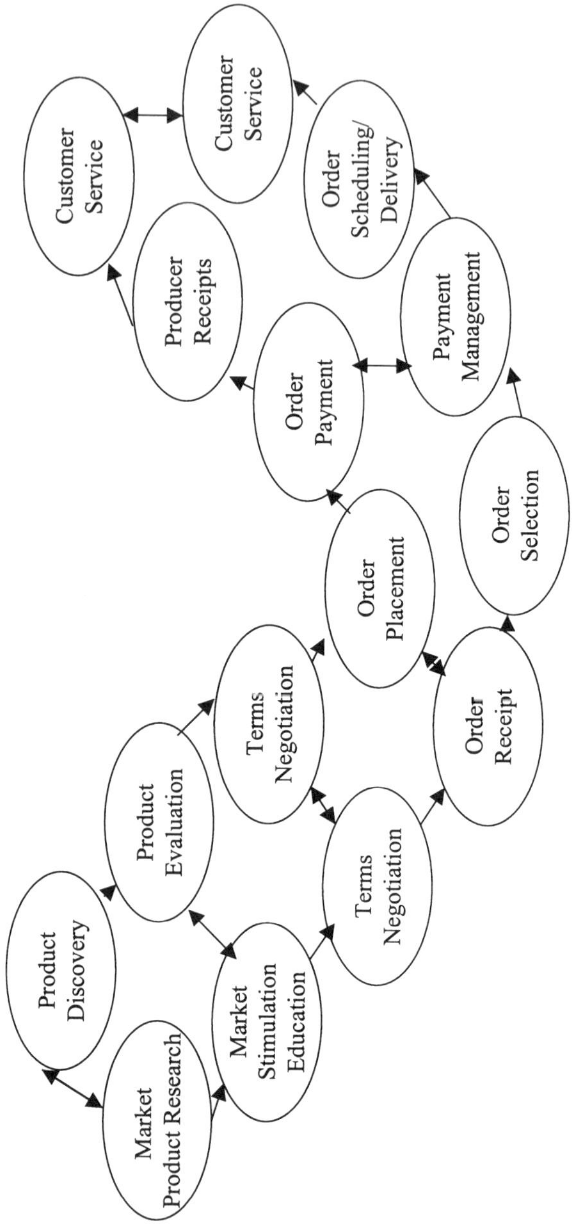

The order of value enhancing activities and degree of interactivity between producers and consumers is very different from the Porter value chain. Because a significant component of E-Commerce value is derived from speed in executing transactions, the first four components of the producer and consumer value branches have direct linkages. First, technology associated with possible products and customer needs is the fundamental driver of the E-Commerce value chain (E-CVC) (Technology Development is a support function in the Porter Model). The second E-CVC node considers the important task of market education and consumer product evaluation. Because of the Internet, this information can be disseminated quickly and can have sufficient content to be evaluated. This information link replaces some of the physical sales and marketing links in the Porter value chain. Much like the Porter value chain, the third and fourth nodes, "term negotiating" and "order placement/receipt," will differ by industry. Whether the consumer or producer dictates the terms of the agreement will depend on who possesses the most power that can be derived from traditional bargaining power models. Interestingly, the availability of information on the Internet can in fact change the relative balance of power between consumers and producer. For example, just a few years ago a car salesman had the information advantage over the buyer, but today Internet based services have turned car purchases into a bid market and given the advantage to consumers. The E-CVC greatly simplifies the order placement and payment process nodes and permits electronic systems to replace paper and human interfaces, thereby significantly reducing cost, decreasing payment time and increasing accuracy of the process. With the exception of electronic tracking and delivery of goods, the final two value chain nodes, i.e., delivery and customer service, perform much the same function in the Porter and E-CVC value chains.

4.4 Information Value Chain Processes

The two main foci of electronic commerce are business-to-business (B2B) transactions and business to consumer (B2C) applications. A study by IBM's E-Commerce Research Lab (1998) indicates that B2B is increasing in importance and three distinct areas are emerging: First is the segment characterized by auctions, bidding and negotiating processes, the second is Internet-EDI and the third is the integration of back-end business processes.

The Internet greatly facilitates auction/negotiating/bidding processes and, in effect, allows a "SPOT" market to be built for each process. Auctions bring together multiple prospective buyers who are electronically linked via the Internet to a real time interactive sale process. The bid management process is asynchronous in nature and involves the procurement/selling of tangible goods. Specifications for these goods are posted on the Internet and bids are then transmitted back for the offerings company's review and decision. Two-party negotiations have information transmitted electronically between two distant locations. A physical record of the negotiation process (e-mail messages) is available for both parties to review. Internet-EDI opens up interfirm connectivity and extends the business process. No longer are costly investment in restrictive private networks required in order for firms to experience the benefits associated with industry trading groups. The lowering of the cost threshold to participate will greatly increase the competitiveness of Internet-EDI markets. The last of B2B applications deals with back-end connectivity where standardized order and inventory measurement systems will greatly facilitate the tracking of interfirm transactions. The net result of the above electronic possesses could be less internalization and that, in fact, electronic markets aid rather than retard inter-firm transactions.

The B2C applications all relate to the very basic marketing saying "know your customer and keep the profitable ones happy." In varying degrees, technology exists to target electronic coupons, patronage awards, cash back offers, free trial/samples, cross sales and contests. Once sufficient data is accumulated via the electronic media, the holy grail of "one-to-one marketing" can be achieved.

In the emerging electronic commerce environment new models of buying and selling are evolving which effectively decouple the information flow with logistics and distribution. Fixed priced models are being replaced with competitive bidding and exchanges /brokerage functions are now dealing with matchmaker rather than staid historical relationships.

4.5 Information Value Chain Applications

Just a few years ago when Internet electronic commerce first appeared on the scene, it promised the world: a competitive edge, new ways of doing business and generous profits (Engler, 1997). Not surprisingly with the early systems there were a few successes and many more failures. Major problems included application complexity, data compatibility, transaction processing, system integration and scalability issues, large-scale data warehousing, and ongoing maintenance and security problems. Fortunately much is learned through trial and error. Today's second generation electronic commerce systems are more focused and center on ten basic characteristics: allows the business to do something it cannot do in the physical world; opens up back-end systems to new constituents; makes the Internet a business critical delivery system, aggregates information; creates a new revenue generating service; chains business transactions to one another; includes adaptive navigation - allowing users to switch mid stream between parametric or concept-centric searching; organizes buyers and sellers; uses registry services or 'matchmaking'; makes interface personalized, dynamic, user friendly and engaging; and can be cost justified.

What appears to be happening now is that electronic commerce is creating a value chain on an industry by industry basis. In book distribution, *amazon.com* took on the industry by offering convenience, lower prices and customization and, not surprisingly, captured a large share of the market with minimum resistant. In other industries, the rate of penetration may be slower; however the end game will likely be the same for those who do not adapt and substitute electronic for physical connectivity and delivery. The rate of adoption of electronic commerce will likely be rapid because of the burdensome cost structure associated with physical linkages of established business. Once electronic commerce establishes low cost inroads and pirates key sectors of the value chain, wholesale abandonment of existing business practices will occur and industries will remain in turmoil until a new value chain grounded in electronic commerce coalesces. Michael Dell has used technology and information to blur the traditional boundaries in the value chain among suppliers, manufacturers and end users. His elimination of the distribution function has allowed him to create a $12B corporation in thirteen years. His virtually integrated corporation offers the advantage of a tightly coupled supply chain that has traditionally come from vertical integration with the benefits of focus and specialization that drives the virtual corporation (Magretta, 1998). In effect, virtual integration means that you basically put together a business with partners that are treated as if they are inside the company.

Though the Internet is getting most of the press coverage, an increasing number of companies are realizing the benefit of intranets which use the same technology as the Internet and can be deployed with relatively little incremental investment (Allen, 1997). (Netscape Communications Corp, the leading provider of browser technology, reports that two-thirds of the world's 240,000 servers are dedicated to internal private use and that will rise to 10:1 by the year 2000.) Firm-specific intranets achieve significant saving by eliminating paper and replacing highly structured data interfaces with a more fluid, inquiry-based approach. Intranets are initially used for enhancing communication (vertical and horizontal), then coordinating activities and decision making and finally for supporting true collaboration among people addressing current issues or planning for the future via virtual teams. They encourage a departure from hierarchies to a more networked culture. Employees are stimulated to initiate action rather than to ask for authorization; value is demonstrated not by withholding information but rather by sharing it (Allen, 1997). General business factors stimulating intranets include: geographic dispersion fueled by mergers and acquisitions and a desire to be closer to customers no matter what part of the world they are in; speed, the new universal phenomenon in business; and the enormous downsizings that have occurred across all industries, which have resulted in fewer people left who know the answers and a greater requirement for cross-departmental coordination (Allen, 1997). As the value chain indicates, a business is not an island and external linkages are critical for success. Electronic commerce facilitates and improves upon these business to business (B2B) and business to customer (B2C) interfaces by the creation of extranets that combine the intranet with the Internet. No longer are designated people within the corporation (sales, purchasing, customer service) the sole interface with the outside world of consumers and suppliers. Firms both reduce the cost and increase the value of these external interactions by allowing customers and trading partners to support themselves by downloading new releases of products, looking up information in online catalogs, querying shipping departments on expected dates, or accounting department on billing issues (Allen, 1997). Federal Express's dial-up website, which permits the real-time tracking of customers' packages, is a good example of an extranet.

It appears that Electronic Commerce has become established as an important business medium of exchange. One major hurdle left is whether the Internet provides sufficient threat for institutions, so that they may impede its progress. Zeller (1997) states that the fundamental question is "How do take a borderless medium like the Internet and interface it with one of the most structured institutions, the tax system?" The stakes are high: a high tax burden would certainly slow growth, be difficult to administer, handicap Internet selling versus mail-order and cause investment shifts much like tax and tariffs have done in the past. On the other side, state and local officials fear a collapse of their tax revenues. So far, Massachusetts and New York have eliminated taxes on Internet sales for five years. Similar moratorium bills are being consider at the Federal level.

4.6 Can electronic commerce be incorporated into existing theories of the firm or are new theories required?

Two of the most quoted theories of the firm are the Resource-Based view put forth by Barney (1991) and that of the Transaction Cost View originally discussed by Coase (1937) and expanded by Williamson (1975). A brief summary of each theory is provided followed by an analysis of the research question.

Resource Based View, essentially concerned with a firm's ownership specific advantages, states the firm's resources are heterogeneous and immobile and provide the firm sustainable competitive advantage over time. Resources are valuable if:

1. they can be exploited or used to minimize competitive threats;
2. they are rare, i.e. a firm's competitor does not possess them;
3. they cannot be imitated or copied perfectly;
4. they are not substitutable, i.e. competitors cannot use similar resources for the same purpose.

Barney (199x) states that resources are everything within a firm's control including information, knowledge, processes, physical and human capital, and organizational structures.

Transaction Cost Economics, essentially concerned with a firm's internalization specific advantages, focuses on the governance of contractual relationships. Williamson (1975) views the market and firms as alternative methods of coordinating production. The firm exists to reduce transaction costs and these costs are, to a large extent determined by the underlying institutional environment, i.e. the set of fundamental political, social, and legal rules that establish the basis for production, exchange and distribution. More specifically, the market hierarchy tradeoff is determined by the relative values associated with the degree of asset specificity, uncertainty, opportunism and number of transactions. Most writing in this area focuses on minimizing cost, but in reality maximizing value (the difference between transactions cost and benefits) determines the efficient organizational modality.

The dynamic version of the Resource Based View can be readily applied to the information age. Resource Based economists would likely argue the use of a new medium to conduct business, such as electronic commerce, represents a rare skill to the firm and, as such, the firm should be able to earn above normal returns on that skill. The same would also be justifiably argued for the new types of relationships that have to be developed to take advantage of the economies associated with the information age. What is new, however, are the processes and applications necessary to take advantage of an information economy. Once acquired by the firm (networked activity set) these become like any physical asset and are used to further differentiate the firm from its competitors.

The application of transaction cost theory is more problematic because the information economy changes the fundamental political, social and legal rules that establish the basis for production, exchange and distribution. Even after the new institutions are established, the relative importance of four hierarchal tradeoff criteria must then be reevaluated to determine their appropriateness.

5. ELECTRONIC COMMERCE'S IMPACT ON TRADE/FOREIGN DIRECT INVESTMENT AND STRATEGIC ALLIANCES

Cerf (1995) stated that one of the radical and exhilarating aspects of the Internet is that it is truly an egalitarian technology. Any company or organization, from five to 500,00 employees, can create and international presence and a new and more productive paradigm for generating future wealth by driving this prototype on the Information Superhighway. We now explore how they may do this.

5.1 Trade

International trade occurs when a buyer in one part of the world is aware of a seller's product somewhere else and both can financially benefit from trade. They both must understand the terms and conditions of the transaction, be able to make payment in a mutually agreeable manner, and arrange for delivery and shipment. The Internet can be used at points on the above value chain to make it more efficient. However, cultural and institutional barriers will initially reduce potential gains.

Adams (1994) states that regardless of how electronic commerce is implemented, the international trading benefits are enormous. A company can eliminate or reduce ad hoc queries regarding its shipping and production schedules by posting that information with its trading partners. In addition, electronic commerce is a very good vehicle for checking the status of orders, communicating back order alternatives, or qualifying order conditions. This lowers telecommunication costs, frees up employees' time and reduces overall friction in the transaction process. The Canadian Government's initiative to electronically standardize customs documentation has been very successful, e.g., Honda Canada calculated that it saved $700,000 in the first year alone (Chudleigh and Sheppard, 1995).

Manchester (1997) quotes Peter Hill of CSC Index Research who stated that the economics associated with Web financial transactions are irresistible. For example, the average cost of a transaction over the home shopping television channel is $3 to $4, whereas electronic shopping over the Internet can be as low as 15 to 20 cents. He also cited a Price Waterhouse estimate that $500M of goods and services was sold over the

Internet in 1995. While this represents approximately .5% of the total credit card sales of $500bn, it is a significant part of the business world and will increase rapidly when institutions solve the security problems. Loeb (1996) states that MasterCard and Visa serve a combined total of about 700 million customers worldwide. Because of these companies' ability to draw banks into the arena, many people see Secure Electronic Transactions (SET) as the catalyst that will finally cause Internet commerce to take off in a big way.

Waters (1997) cites one example of an auto dealer who appeared on the Web last summer and now sells about 10% of its cars that way. In 1997, only about 5-10% of the dealers are on the Web; however, this is expected to increase dramatically in the next few years year (Wymbs, 1997). One result could be the emergence of super dealers attacking the somewhat cumbersome distribution system for the $300bn vehicle market. However, the next logical step could be for manufacturers to directly connect with consumers and bypass dealers entirely. Customers could get custom designed products more quickly and at a lower price. Manufacturers could gain a wealth of market information about their customers and the dealers' roles could be dramatically reduced, maybe only to servicing.

Electronic commerce will assist trade for virtually any product or service that exhibits match- making characteristics. The wider the cultural or information asymmetry between the markets, the greater the opportunity for electronic commerce to result in a successful transaction. In many cases, the Internet becomes the unifying platform, which permits information access, scanning and use by both parties.

5.2 Foreign Direct Investment (FDI)

FDI occurs when a firm can economically exploit its ownership advantages, internalization capabilities and/or specific locational parameters over its domestic counterparts (Dunning, 1993). The main types of foreign investment are: resource seeking, market seeking, efficiency seeking and strategic asset acquiring. In effect, FDI occurs because a firm believes a hierarchal approach will yield greater profits than a marketplace solution. Any items that enhance a firm's ability to communicate or to obtain a better knowledge of the market vis-a-vis its competitors will encourage FDI, while those that make the market (vs. the firm hierarchy) relatively more efficient will discourage FDI. The vast majority of information available via Internet facilitates global scanning and the identification of target companies for strategic asset seeking investment, *i.e.*, acquisitions.

Over the last year, Shafe (1997) found that the Internet has become the number one technology for gaining competitive advantage. There are increasingly new applications for business change arising from thc Internet that were undreamed of just a year ago. AT&T is building internal Internet service that can be used by 30,000 employees to give them unprecedented competitive advantage. Using client/server software, it would have taken AT&T two years to build this network, but by using the Internet, AT&T was able to build the network in weeks. This clearly enhances a firm's ownership advantage and encourages FDI.

Adam and Yesha (1997) state that data mining can lead to a company processing more, and gaining a better understanding, of its information. By exploiting these informational relationships a company obtains better decision-making capabilities and ultimately achieves greater competitive advantage and improved decision-making. Similarly, the driver behind push technologies in the business world is gaining competitive advantage. Naim (1997) predicts that push technology, which allows a publisher to distribute tailored content directly to a consumer's desk, will account for $5.7bn, more than one third of the revenues generated on the Internet in the year 2000. The first and best-known push vendor is Pointcast, which launched its products in early 1996. Selling points include reduced frustration by executives exploring the Internet and more efficient information collection.

With regard to market seeking investment, the Internet affords a firm quite unique intelligence gathering capabilities, *e.g.*, mailing lists, news groups, etc. E-mail can also be effectively used with customers because communication takes place in writing, therefore a record of all transactions is maintained. By understanding marketing and sales practices, businesses can identify their target customers, understand technical capabilities of these products and use the Internet to enhance sales performance (Maloff, 1997). Maloff also believes that a global request for proposal (RFP) bid process through appropriate news groups and other Internet forums will grow in importance. After relationships are established and winners selected, both sides will then seek ways to efficiently use electronic communications for their daily interactions.

Alternatively, global connectivity reduces the need for a firm to internalize transactions and makes the electronic market much more efficient. Maloff (1997) stated that now organizations can create virtual extensions of themselves, domestically and internationally, for relatively reasonable and predictable costs. In effect, every server can be viewed as a virtual storefront without any FDI at all. Therefore, the overall effect of FDI on electronic commerce is ambiguous.

5.3 Alliances

According to Dunning (1995), the choice between a hierarchal (FDI) and alliance (shared) modality as a means of lessening arm-length market failure clearly depends on their respective costs and benefits. In effect the choice rests on a trade-off between the perceived benefits of sharing risks and capital outlays on the one hand and the costs of loss of control associated with reduced (or no) ownership on the other. Hagedoorn (1993) found that the objectives of most strategic alliances have been to gain access to new and complimentary technologies, to expedite innovation and to enhance the efficiency of a particular function, *e.g.*, R&D. In effect, strategic alliances are formal arrangements of two or more firms that result in cooperation in the provision of a joint activity. The Internet will certainly permit the more efficient screening of candidate firms around the world, a necessary condition. However, the critical sufficient conditions are inter-cultural, personal interaction and negotiation, which suggests limitations (boundaries) on the Internet (electronic commerce).

Adam and Yesha (1997) describe the matching-making concept associated with trade; however, it is equally applicable to finding alliance partners. Conceptually, the process involves firms electronically seeking access to products, markets, and/or capabilities of other firms that through cooperation could result in synergistic benefits to both firms. Historically companies have limited their supplier network to firms that are local or known to them. The Internet effectively removes this constraint and permits the firm to seek suppliers and other vertical alliances for standardized products around the world. The removal of distance sensitivity associated with electronic commerce will help bring together horizontal alliances associated with access to complimentary technology and networks of similar firms. Electronic commerce is also becoming a great size equalizer. Cerf (1995) points out that the Internet is already providing small firms with volumes of previously inaccessible data and business intelligence through these networks. Small firms can form virtual teams or outright partnerships to do business on a global scale that rivals their larger competition - all without the expense of outpost offices. Therefore, the overall effect of electronic commerce on strategic alliances will likely be positive.

The bottom line effect of electronic commerce is that the decision maker will have more information and greater choice regarding organizational modalities available to maximize profits in an increasingly competitive marketplace.

With the effect of electronic commerce identified for each global transaction activity, now an attempt is made to address specific comparative questions. These answers are hoped to be directionally accurate, but certainly will not be precise given the dynamic, unpredictable path electronic commerce will take over the next several of years.

5.4 Will there be less need for FDI and strategic alliances if international trade is made more efficient through Internet information and financial transfer systems?

There is little doubt that the Internet will make international trade more efficient. Adams (1994) states that electronic commerce, because of its ability to easily share a wide range of correspondence and data with trading partners, is the solution to communicating with overseas trading partners. Similarly, Wheelwright (1997) puts forth a positive trade hypothesis and states that the biggest potential for Internet trade lies in goods that can be ordered, paid for, processed and delivered digitally. They include computer software, magazines, newspapers, movies, music and specialized video programs.

Given increased trade efficiency, however it is not clear that there will be less need for FDI and alliances. According to Denton (1997), the Internet is making it much easier to reach the scale of production that leads to adequate rates of return. By making it easier to distribute news and software, it is creating a seamless global market for many information technology products, thereby reducing the amount of capital required for international ventures. The matchmaking powers of the Internet also positively affect FDI and alliance activities. Hart (1996) poses the question: "Looking to find a business partner for your small business or home-based company?" and proceeds to answer it: "Just as it has become possible to find employees online, the Internet has become an effective medium for bringing together people around the world as business partners -- people who probably would otherwise never meet." He further asserts that people are more accessible on the Internet. But if partners are going to invest money together, they must trust each other, and you cannot trust a stranger on the Internet. Not even a virtual meeting via Internet real-time video is a viable substitute for a face-to-face meeting.

Also, increased trade will bring new pressures on hierarchical and alliance modalities, forcing them to innovate and become more efficient. Mullich (1997) makes the observation that with competitive pressures increasing as quickly as user demand, players are forging a host of partnerships to offer value-added services. It is likely that in most industries all three (trade, FDI, alliances) will coexist; however, one will likely dominate based on the unique industry characteristics, e.g., alliances in telecommunications, FDI in automotive and trade in manufacturing areas.

5.5 Will the Internet, through the removal of time and space boundaries, have its greatest initial impact within firms and cause dramatic increases in intra-firm trade and FDI?

Petrillo (1995) stated that his company's (AT&T) free phone service "800 numbers" was the first form of electronic commerce and partially removed time and space boundaries. He went on to say that the Internet was at about the same stage today that free phone service "800 numbers" was in the early 1980s in the United States. Today, AT&T's network handles 15 billion free phone calls per year. The Internet has already become instrumental in linking most parts of the value chain both within and between firms. Cerf (1995) estimated that 40% of all U.S. manufacturers were electronically linked to their suppliers, partners and customers. This creates not only greater efficiency but greater bonds. With regard to Intranets, he used Chrysler as an example of what is possible. Chrysler engineers automobiles around the clock by sending information back and forth through the Intranet among work sites in Detroit, Europe and Asia.

Similarly, Black (1997) stated that Intranets could deliver spectacular returns on a relatively small investment in a short period of time. He cited an International Data Corporation (IDC) study that found greater than 1000% returns for Intranets built on Netscape software with a payback of six to twelve weeks. A substantial part of the savings comes from the elimination of storage, printing and distribution of documents that are made available online. Because a firm can exercise greater controls on internal standards, Intranets are a short-term winner. Shillingford (1997) boldly predicted that Intranets would be a $50bn market worldwide by the end of the decade. Shillingford cited airlines and aerospace companies cutting costs by sharing digitized information from manufacturers via a single connection to Sita's Aerospace Intranet rather than through expensive international leased lines. Mr. Santiago of GE Information Services, the leader in electronic commerce, believes it will take two years for the Internet to become stable, so there will be an evolution from virtual private networks to linked Intranets and then to the Internet.

What we see are many divergent projections of a service that is in its early life cycle phase. It is clear that Intranets can more easily be managed than the Internet and it is also clear that the Internet is a more cost effective long run solution to most global commerce opportunities. The timing of this transition is not clear because it is dependent on economic, political and sociological agreements among many nations of different cultures and beliefs.

6. CONCLUSION

This analysis has addressed the impact of the information revolution on the global corporation. Because the paradigm is in its initial phase, the best one can hope for is to be directionally correct in one's conclusions. However, the main purpose of this analysis is to stimulate thinking on where the Internet, the overarching platform for electronic commerce, is heading and how it will affect global commerce. The parameters central to the analysis, i.e., the value chain, trade, FDI and alliances, will remain important throughout the paradigm evolution. The drivers of these parameters will become better defined with time, and trade-offs among and between them will eventually be quantified and refined.

Chapter 10

Privacy, Property and Policy

Hidden Implications for the Information Highway

VALERIE STEEVES
Carleton University

Contrary to popular wisdom, I have come to believe that technology is not neutral. Technology is a choice, a way of doing something. And that choice cannot be isolated from its social and political consequences. This was true of the wheel, the wagon, the telegraph, the railroad. And it is true of the Internet.

Moreover, I have also come to believe that we have always lived in an information society. Human nature demands that we learn and grow. To do so, we acquire information, and we call the totality of our experience in that endeavour knowledge. All human societies have been based upon collective knowledge, from the first farming communities to the manufacturing plants of Dickens' London to the online communities of the Internet.

So as we seek to navigate the information highway, we have a wealth of knowledge to draw upon. The trouble is, much of that knowledge reflects socially acceptable forms of exploitation and privilege. Do we really want to move to the "information society"? Or would we rather entrench that privilege in spite of vast social pressures for reform? 'Reform': re-making; the alchemist's art of taking the pieces of what is and creating something completely new. Perhaps even golden.

1. INTRODUCTION

According to Moore's Law, computing power doubles every eighteen months. Law, on the other hand, evolves over centuries, so it is no surprise that our legal system has found it difficult to keep pace with new technologies. However the conflicts which have arisen because of technological innovation are not just the result of conflicting velocities. In many ways, they are symptomatic of a deeper dynamic, a fundamental re-ordering of power relationships within society as a whole. As this re-ordering has challenged the pre-eminence of vested interests, the people who represent these interests have attempted to use the law to retrench their rights, in spite of changing social and political realities.

This paper will examine the ways in which networked communications are challenging our existing understanding of two central legal concepts: privacy and property[1]. Both privacy and property are longstanding concepts designed to protect the integrity of the individual and the creative process. However, the laws governing these concepts are rooted in the political economy of the Middle Ages, and as such, seek to protect the relatively weak individual from the invasive all-powerful state. The online world lacks the hierarchical structure of the traditional nation-state, and is based on very different paradigms or concepts of community and information sharing. These new paradigms have been developed outside the traditional vehicles for policymaking, and are often in conflict with established legal interests.

At the same time, the non-hierarchical structure of the Internet provides the state and the private sector with unprecedented opportunities to invade the privacy of the individual for economic gain. The growing practice of tracking the online actions of consumers, employees and citizens is often inconsistent with our core legal notions of democracy, individual freedom and human dignity.

In this paper, I will analyze the socio-political rationale underlying both privacy and property, and determine whether or not current laws and policies can effect an appropriate balance between technologically-enabled efficiencies and individual autonomy. I will also suggest ways in which the legal framework protecting privacy and property can be reformed to better balance the benefits of networking technologies against the social and political values at the root of a healthy democracy.

In order to provide some context for this discussion, I begin by examining one example of the current conflict over online property and privacy, the MP3 controversy.

2. ONLINE MUSIC - AN ILLUSTRATION OF EMERGING CONFLICTS

Early in 1999, the rap group Public Enemy posted one of their songs on their web site using MP3 compression software.[2] Public Enemy fans were able to listen to a new single, the band was able to promote its upcoming tour and everyone was happy - everyone except Public Enemy's recording label. Unlike small labels, which actively encourage their artists to use MP3 to promote their songs, the distributor of Public Enemy's music argued that this was an infringement of their intellectual property rights and threatened to sue the band unless they removed the file from the Net.

Legal battles between artists and publishers are not new. However, this is not a tidy battle over technical rights. The MP3 case underlines the fact that intellectual property laws are designed to create monopolies. In a world where recording companies control access to the infrastructure necessary to record and distribute music, artists have little bargaining power when it comes to getting their work out into the public domain unless they bargain away the lion-share of their rights. But in a networked society, where an artist can circumvent the normal channels of distribution by directly communicating with fans, legal monopolies become harder to enforce because the holder of the legal monopoly is no longer in control of the means of production or distribution.

The MP3 controversy has also demonstrated that the number of people able to encroach on legal monopolies grows exponentially in a networked world. In order to maintain a monopoly over distribution in such an environment, the holder of the monopoly must seek to control the actions of an ever larger group of people. Public Enemy's recording label, for example, did not just threaten to sue the band; it also threatened to sue the online search engine Lycos because it offers a quick search function of all MP3 sites on the web.[3] In a similar move, the Recording Industry Association of America filed a suit in the American courts in October of 1998 to prohibit individuals from using a new pager-sized device called the Rio to download, store and play MP3 files.[4]

To truly control distribution of MP3 files, the recording industry will be forced to expand its efforts to encompass individual acts of consumption. However, controlling the actions of the online community has not proved to be an easy task. The German branch of the International Federation of the Phonographic Industry recently released a statement that it has shut down over 100 web sites using MP3, claiming that the use of MP3 cost the music industry at least 20 million marks ($10.7 million) in 1998[5]. Nonetheless, the downloading of MP3 files over the Internet continues to explode, and not just among unlawful rebels at the fringe of society. The activity on the

MP3.com's web site, for example, peeks at noon, falls off at 5 o'clock and drops significantly on the weekends. In other words, usage mirrors the work day and MP3.com's best information indicates that its average visitor is also the average office worker and executive.[6] Given the mainstream use of the technology, MP3 supporters argue that the music industry would be better served by developing other ways to make money, such as creating their own MP3 sites which could be supported by advertising and subscriptions.[7]

So far, the music industry has been slow to warm to this advice. Instead, it has attempted to broaden its legal reach and control a growing range of personal activity. Ironically, it has been assisted in this task by the same networking technology that provides music fans with a free distribution system. Prior to the advent of the Net, for example, people commonly copied and exchanged the lyrics and guitar chords to popular songs. However, once those individuals began to do the same thing over the Internet, the recording industry began to locate them and take legal action to stop them.[8]

One could argue that the industry is merely attempting to stop the unauthorized reproduction of their intellectual property.[9] However, in order to enforce its legal rights, the industry is using the open architecture of the Internet to monitor personal activities which, prior to networking, were clearly within the private realm.[10] The Net, then, is a two-edged sword. As much as it provides individuals with a means to distribute information, like song lyrics, it also gives the holders of legal monopolies a way of placing individuals under surveillance in order to control their use of the information they find there.

3. PRIVACY AS POLITICAL RIGHT

The use of intellectual property laws to invade personal privacy for the purpose of exercising control over the private actions of individuals is a profoundly disturbing one. Privacy has long been recognized as one of the hallmarks of a free and democratic society. Mr. Justice La Forest of the Supreme Court of Canada wrote that privacy:

> ... is at the heart of liberty in a modern state. Grounded in man's physical and moral autonomy, privacy is essential for the well being of the individual. For this reason alone, it is worthy of constitutional protection, but it also has profound significance for the public order. The restraints imposed on government to pry into the lives of the citizen go to the essence of a democratic state.[11]

The Supreme Court has interpreted section 8 of the Canadian Charter of Rights and Freedoms, which guarantees the right to be free from unreasonable search and seizure, to include a right to privacy.[12]

Our understanding of what that right to privacy means is rooted in the history of the common law. Traditionally, a liberal democracy seeks to protect the individual from the overwhelming power of the state. The Magna Carta, for example, is considered to be a seminal document in which the King relinquished absolute power and first subjected his actions to the rule of law. The most oft-quoted paragraph of the Magna Carta reads as follows:

> No freemen shall be taken or imprisoned or disseised or exiled or in any way destroyed, nor will we go upon him nor send upon him, except by the lawful judgment of his peers or by the law of the land.[13]

However, almost all of the remaining sixty-eight paragraphs in the document laid down strict limits on the King's power to seize the Barons' property. In a feudal society where the Crown owns virtually everything, the protection of private property was a political revolution.

Both privacy and property, then, were first articulated as political values. Today, almost eight hundred years later, the state's ability to invade our homes and seize our physical property is still strictly limited.[14] However, both the state and the private sector are now collecting and selling our personal information at unprecedented rates. Federal Privacy Commissioner Bruce Phillips warns us that:

> We are in fact buying and selling large elements of our human personae. The traffic in human information now is immense. There is almost nothing the commercial and governmental world is not anxious to find out about us as individuals.[15]

The courts have been slow to respond to the changing dynamics of the electronic trade in personal information. In shaping our constitutional right to privacy, the Supreme Court has focussed on those circumstances in which people have a reasonable expectation of being secure against a search.[16] This is problematic for an open environment like the Internet where all communications traffic travels over publically accessible cables. Moreover, the factors the Court has identified in determining whether or not the reasonable expectation test is met include: presence at the time of the search; the ability to regulate access; and the objective reasonableness of the expectation.[17] The use of cookies and other tracking software that can search through an individual's hard drive every time they visit a web site makes it difficult to argue that he or she has a reasonable expectation of privacy. The user is present at the time of the search, can choose to avoid sites that use invasive technologies, and is knowingly communicating over a public network. In other words, the architecture of the Internet provides people with no de facto expectation of privacy at all.[18]

The Court has also been slow to grasp the ways in which electronic information is used. Data matching and data mining software enable organizations to pool their data and construct detailed profiles on the private lives of individuals.[19] Privacy invasions that flow from these practices are not the result of the acquisition of each particular piece of information. Rather, they come about when these disparate bits of data are linked together.

This was precisely the point the Supreme Court failed to appreciate in the case of R. v. Plante.[20] The police received an anonymous tip that Mr. Plante was growing marijuana in his home. Since they had no evidence to support the allegation, they were unable to obtain a warrant to search the premises. Instead, they reasoned that indoor cultivation would increase Mr. Plante's electrical consumption. To test their theory, they logged in to the city utility's mainframe and searched Mr. Plante's electric bills. The Court argued that this was not an invasion of Mr. Plante's privacy. In the words of Mr. Justice Sopinka:

Section 8 of the Charter should seek to protect a biographical core of personal information which individuals in a free and democratic society would wish to maintain and control from dissemination to the state. This would include information that tends to reveal intimate details of the lifestyle and personal choices of the individual. The computer records in this case, while revealing the pattern of electricity consumption, could not reasonably be said to reveal intimate details of the accused's life.[21]

As more of our activities migrate to the Internet, it will become ever easier to collect disparate pieces of information about us as individuals: the books we buy; the bank we use; the records we listen to. Alone, none of these pieces of information reveal much about our biographical cores. But linked together, they provide others - invisible others - with detailed and revealing portraits of our personal lives which can be manipulated and exchanged for profit. Privacy, like property, has gone from being a political right, to an economic one.

4. PRIVACY AS COMMODITY

Existing privacy legislation does little to discourage the commercial exploitation of personal information. Although every jurisdiction in Canada excluding Prince Edward Island has laws that regulate the ways in which governments collect and use personal information[22], only Quebec has extended those laws to cover the private sector[23]. The remaining provinces have a patchwork of voluntary codes and sectoral practices. Concerned about the uneven playing field this leaves for the private sector and a growing sense of distrust among consumers[24], the Canadian Standards Council of Canada attempted to fill the void in 1996 by establishing a national standard in its Model Code for the Protection of Personal Information[25].

The Model Code is the result of negotiations between government, industry and consumer groups, and sets out ten principles of fair information practices:

1. *Accountability:* An organization is responsible for personal information under its control, and shall designate an individual or individuals who are accountable for the organization's compliance with the Code's principles.
2. *Identifying Purposes:* The purposes for which personal information is collected shall be identified by the organization at or before the time the information is collection.
3. *Consent:* The knowledge and consent of the individual are required for the collection, use or disclosure of personal information, except where appropriate.
4. *Limiting collection:* The collection of personal information shall be limited to what is necessary for the purposes identified by the organization. Information shall be collected by fair and lawful means.
5. *Limiting Use, Disclosure and Retention:* Personal information shall not be used or disclosed for purposes other than those for which it was collected, except with the consent of the individual or as required by law. Personal information shall be retained only as long as necessary for the fulfilment of those purposes.
6. *Accuracy:* Personal information shall be as accurate, complete and up-to-date as is necessary for the purposes for which it is to be used.
7. *Safeguards:* Personal information shall be protected by security safeguards appropriate to the sensitivity of the information.
8. *Openness:* An organization shall make readily available to individuals specific information about its policies and practices relating to the management of personal information.
9. *Individual Access:* Upon request, an individual shall be informed of the existence, use and disclosure of his or her personal information, and shall be given access to that information. An individual shall be able to challenge the accuracy and completeness of the information and have it amended as appropriate.
10. *Challenging Compliance:* An individual shall be able to address a challenge concerning compliance with the above principles to the

designated individual or individual s accountable for the organization's compliance.[26]

The need to obtain the consent of the individual before his or her personal information can be collected is a powerful one, but it is the exception to the rule, the cases where obtaining consent or knowledge is "not appropriate", which will determine the effectiveness of the CSA Standard.. Unfortunately, there is nothing in the Standard that defines these cases, or allows the citizen to question whether or not the information should be collected. As long as the organization identifies its purpose, then it can use the information for that purpose, even though such use may have consequences that are detrimental to the individual or society as a whole.

Insurance is the best example of this potential for abuse. The social good behind the concept of insurance is that the risk of harm to an individual is, through the insurance process, distributed over society as a whole. Insurance protects each of us from the financial ramifications of unforeseen injury by spreading the cost of that injury over all policyholders. However, the insurance industry already discriminates against individuals on the basis of health and personal choice. For example, juvenile diabetics cannot buy life insurance because of their pre-existing health condition, and smokers are required to pay a premium because of their increased risk of health problems. As the information industry gathers more of our personal data -- our lifestyle choices, our genetic predispositions to possible health problems in the future -- we run the risk of broadening that discrimination. In 1997, two pregnant women in New Brunswick who faced the possibility of delivering children with disabilities were instructed to submit to psychiatric evaluation when they refused to undergo genetic fetal testing.[27] And an American health insurer refused to pay the medical costs for a baby born with Down's syndrome because the mother had refused to have an abortion after genetic screening indicated the child was at risk for the disease. In each of these cases, the insurer was seeking to limit its financial risk. But there are greater social issues at play. The decision to deny access to financial services to classes of people or to eradicate genetic illnesses by eradicating the people who have the illnesses is not a question of fair information practices; it is a question of ethics and social responsibility. The Model Code is ill equipped to help us deal with these issues, because it presupposes that the commercial flow of personal data should and will continue.

In October 1998, the federal government introduced Bill C-54, the Personal Information Protection and Electronic Documents Act, which applies the Model Code to all federal works and commercial activities. Bill C-54 was designed to promote the use of personal information as a commodity in the information marketplace.[28] In the discussion paper that preceded the Bill, the federal government argued:

> Legislation that strikes the right balance between the business need to gather, store, and use personal information and the consumer need to be informed about how that information will be used ... is an important element of building the consumer trust and the market certainty needed to make Canada a world leader in electronic commerce [emphasis added].[29]
>
> [and]
>
> The ability to provide effective protection for personal information may be crucial to Canada's ability to remain competitive internationally in the global information economy. ... This [European Union] Directive has the potential to make the protection of personal information a major non-tariff trade barrier with Canada ... Canadian businesses may be forced to undertake individual comprehensive negotiations to show compliance with the European Union rules. This process will be fraught with uncertainty and could become lengthy and expensive.[30]

The Bill received the qualified support of many privacy advocates because it was "better than nothing"[31], especially after it was amended to include a purpose limitation clause which limited the collection of personal information to "purposes that a reasonable person would consider are appropriate in the circumstances."[32] However, the business community was divided in its support. The insurance industry, for one, argued strongly that the Bill would detract from its ability to conduct its business.[33] After a long debate in Committee, Bill C-54 died on the order books when the House of Commons prorogued in June 1999.

5. MONOPOLY AND INNOVATION

The desire in Bill C-54 to protect trade is mirrored in the federal government's approach to intellectual property on the Internet. Canada is a signatory to the World Trade Organization (WTO) Agreement on Trade-Related Aspects of Intellectual Property Rights[34] (TRIPS) and the World Intellectual Property Organization's Copyright Treaty[35] (WCT) and Performances and Phonograms Treaty[36] (WPPT). The TRIPS Agreement is the most comprehensive multilateral agreement on intellectual property and provides for uniform standards, enforcement provisions, and dispute settlement mechanisms. All three agreements, especially the WPPT, seek to strengthen existing monopoly rights in the face of new networking technologies.

When the Department of Foreign Affairs and International Trade conducted a series of consultations in April 1999 to canvass the business community's perspective on Canadian trade policy issues in light of its international obligations, it focused on technical issues[37] and assumed that existing intellectual property laws are necessary to protect creative endeavours. For example, the consultation backgrounder states:

> Copyright and neighbouring rights provide the key form of protection to promote cultural industries and cultural activities... Protection is afforded to those who create works, as well as those who contribute to works by interpreting or producing particular versions of works.[38]

The Department also agreed with the WTO's position that intellectual property laws are essential to the development of the information highway. The WTO argues "intellectual property plays an important role (sic) in promoting the development of the infrastructure of communications networks," and "the expected growth of electronic commerce is closely linked with the growing importance of intellectual property".[39]

Intellectual property laws assume that innovation will only flourish in an environment where the innovator is sure to enjoy the economic fruits of his or her labour. In the words of European Single Market Commissioner Mario Monti:

> There must be a determined fight against [online] counterfeiting and piracy since they jeopardise the proper functioning of the Single Market and are prejudicial to the development all forms of creativity and to the growth and competitiveness of European industry. All those involved, inventors, industrialists and consumers, must have a guarantee that the Single Market is a secure environment for their activities and the quality of their purchases.[40]

And yet the history of the Internet tells us that innovation in a networked society flourishes when information is shared rather than monopolized. The infrastructure of the Net was built cooperatively by universities and research institutions. The existence of common protocols enabled those organizations to distribute the costs of infrastructure development. The first backbone of the World Wide Web, MOSAIC, was a freeware browser, and even today, individuals can download commercial browsers for free. When Phil Zimmerman developed the first encryption software, he did not seek to obtain a legal monopoly over the program; he posted it on Usenet so everyone could have a copy. Commercial successes like YAHOO are predicated upon free and unrestricted access by millions of users a day.

Copyright laws, on the other hand, are designed to protect the fruits of a very different kind of communications technology: the printing press. In the oral tradition that preceded the invention of writing, knowledge was a communal asset that was shared by all members of the community. Information was not static, but grew and changed in each retelling of the community's stories. The development of the written word transformed both the method of transmitting information and the nature of information itself as it enabled an author to encapsulate knowledge in a static and unique form.

After the printing press was invented and literacy rates began to rise, a struggle developed between authors and press owners. Over time, press owners asserted their technological monopoly over the means of production and authors were forced to trade away most of their rights to their creative works in order to gain access to the public domain.[41] Copyright laws entrenched the technological monopoly of the press owners by creating a legal monopoly over the right to reproduce or communicate the author's work.[42] In order to be published, authors were, in practice, required to surrender their rights in exchange for royalties.

The situation on the Internet is quite different. Authors now have direct access to the public, and the mass circulation of identical printed copies is slowly being replaced by the transmission on demand of customized information.[43] The technological monopoly of the information distributor no longer exists, and individuals are able to publish and disseminate creative works directly. Not surprisingly, as the story of Public Enemy illustrates, the battle over online copyright is often between creator and distributor.

In spite of the fact that the technological monopoly of distributors has been superseded, governments are seeking to retrench the legal monopoly they enjoy in increasingly intrusive ways. The European Commission, for example, advocates a broad interpretation of the concept of "reproduction" which would include intermediate copying, storage, loading and transmission of files over the Internet.[44] As the Commission's Legal Advisory Board has rightly pointed out:

> ... applying this approach to the digital networked environment will have far-reaching consequences, which the Commission may have overlooked. In the process of being transmitted and routed through a switched digital network, information is constantly being "stored and forwarded". The Commission's approach would imply that practically every act of downloading and screen display would qualify as countless acts of reproduction of the protected work. Thus, for example, sending electronic mail, "browsing" the Internet and viewing a digital file would become restricted acts.[45]

The Legal Advisory Board encourages the Commission to focus not on the technological details of online transmissions but on the "normative essence of the economic right, i.e. making protected works available to the public".[46]

The social utility that flows from the public exchange of knowledge was not lost on the drafters of our copyright laws. The unique technology of the printing press enabled them to protect the written form of a creative work without limiting the communication of the ideas or facts contained within that work. They were also careful to protect the public domain. Copyright protection does not extend to works where the term of the copyright has expired, for example;[47] nor does it prohibit the private use or consumption of protected works.[48]

An information society is predicated upon open access to and exchange of information and knowledge. The economic rights of the creator have never, in law, prohibited acts of private consumption or reception. The data trails created by networked communications have done more than provide an opportunity for monopoly holders to police individual actions; they have enabled copyright holders to claim a monopoly over what, in essence, remain private acts of consumption simply because they take place on a public network.

6. PROTECTING PRIVACY AND PROPERTY IN A NETWORKED WORLD

The link between privacy and intellectual property is not random. The battle between monopoly holders and private citizens taking place today is, in essence, a political one. In the words of the Electronic Frontier Foundation:

> Architecture Is Policy. The architecture and standards for how machines and people interact have implicit implications for social behaviour and human-to-human interactions... [The MP3 controversy] is not just about music. All kinds of information flows in digital audio forms, including talk radio, political speeches, commercial speech such as advertising, private discussions recorded by the participants, telephone calls including conference calls, public meetings and speeches on myriad topics, and spoken books.

Placing the issues in the context of political rights - privacy, freedom of speech, individual autonomy - will enable the courts to better balance the normative values of intellectual property law against technologically-enabled privacy invasions. A recent decision of the 9th U.S. Circuit Court of Appeals in San Francisco emphasizes the importance of upholding the private use of information. The court dismissed the recording industry's allegations that the Rio, a popular device used to download, store and play MP3 files, violates anti-piracy laws. Judge Diarmuid F. O'Scannlain wrote "The Rio's operation is entirely consistent with the [law's] main purpose -- the facilitation of personal use."

Of course, as the European Commission's Legal Advisory Board has pointed out, clarifying intellectual property rights on the Internet by distinguishing between private consumption (which is not prohibited) and public communication of protected works (which is prohibited by anyone other than the holder of the copyright) presupposes "that a clear borderline can be found between 'public' and 'private' communication."[49] In order to define this borderline, we must move beyond the data protection model of privacy protection expressed in current privacy legislation, and articulate a broader definition of personal privacy. To date, that has been done most effectively by the Standing Committee on Human Rights and the Status of Persons with Disabilities in their 1997 report *Privacy: Where Do We Draw the Line?*.[50]

The Standing Committee urged the Government of Canada to adopt a privacy charter which would declare and entrench fundamental privacy rights and the responsibilities attaching to these rights. Specifically, the Charter states:

1.1 Everyone is entitled to expect and enjoy:

-physical, bodily and psychological integrity and privacy;

-privacy of personal information

-freedom from surveillance

-privacy of personal communications, and

-privacy of personal space.

1.2 Everyone is guaranteed that:

-these privacy rights will be respected by others adopting whatever protective measures are most appropriate to do so;

-violations of these privacy rights, will only be allowed if the interference with these rights and guarantees is reasonable and can be demonstrably justified in a free and democratic society.

The basic duties owed to others to ensure their privacy rights are adequately respected include:

-the duty to secure meaningful consent;

-the duty to take all the steps necessary to adequately respect others' privacy rights or, if their rights must be infringed, to interfere with privacy as little as possible;

-the duty to be accountable;

-the duty to be transparent;

-the duty to use and provide access to privacy enhancing technologies;

-the duty to build privacy protection features into technological designs.

The protection of privacy as a human right is consistent with the culture emerging on the Internet, as it clearly places control over the individual's online activities in the hands of the individual. In many ways, the legal tensions between monopoly holders and online users of information are based upon conflicting conceptions of community. Our legal system assumes, correctly or not, that laws grow from the consensus of citizens, expressed through the democratic process. However, the community involvement in this process is a passive one; the power rests with the state, which articulates and enforces blanket prohibitions. The citizen is, in many respects, assumed to be only a passive participant in the legal process and fulfills his or her civic duty by complying with the law.

But our experience of community is very different. Both in the real world and online, communities grow from interpersonal sharing and support -- from the neighbour who looks after your children when you break your leg, to the newsgroup member who passes on a bread making tip. The kind of self-actualizing community rules which grow from online interaction can be censorious, but an offending individual can always choose to flow into another community. Because groups are self-selected, the use of social censure does not preclude participation. It merely sets the limits of how people in that particular group treat each other.

This is not true of the law. The power of the state to forbid certain activity is all encompassing and people who fail to conform to the rules are punished. This power is why the law does not seek to intrude into private activity in all but the most exceptional circumstances. Legal sanctions must be applied thoughtfully in a democracy, because the state's monopoly over the use of force is, by definition, invasive.

Before networked communications, the law did not have entry into this personal aspect of community and citizens were able to set private standards for behaviour. The sharing of knowledge continued to be a communal activity, as it was in the oral society that preceded the printing press. Indeed, it was the line between public distribution and private consumption that protected us from a loss of autonomy and ensured that the normative goal of the law - that information would be made available to the public - was met. However, now that that line is no longer firm, the state's use of blanket prohibitions and monopolies creates a much broader sense of coercive censure and subsequent loss of freedom.

As we move into a networked society, we must be mindful that our legal and technological choices will have social and political consequences. Just as the printing press brought about a decline in traditional, orally-transmitted knowledge[51], the use of monopolistic systems will promote the economic agenda of the few at the price of the free flow of information. We will best be able to meet the challenges associated with the online world if we ground ourselves in the importance of individual autonomy, democratic expression and personal privacy.

Notes

[1]Given the non-tangible nature of online communications, the primary focus of this paper is on intellectual property in general, and copyright law in particular.

[2]MP3 (Motion Picture Experts Group, Audio Layer 3) is software that compresses digitalized audio files, which enables people to download audio files over the Internet with little loss of sound quality.

[3]See Electronic Frontier Foundation, *Digital Audio and Free Expression Policy Statement*, http://www.eff.org/pub/Intellectual_property/Audio/EFF_audio_statement.html .

[4]The suit was dismissed on June 15, 1999 by the 9th U.S. Circuit Court of Appeals in San Francisco. See p. 20 below. "Court approves Internet MP3 music recorders", posted on CNN (http://www.cnn.com) on June 16, 1999 at 7:48 AM EDT (1148 GMT).

[5]See "German MP3 sites shut down", posted on CNN (http://www.cnn.com) on June 11, 1999 at: 3:34 p.m. EDT (1934 GMT)by Mark Lisabeth D'Amico.

[6]Interview on CBC Radio, April 3, 1999 at 3:20 o'clock in the afternoon.

[7]Note 4 *supra.*

[8]In January, 1999, for example, the American National Music Publishers' Association (a trade group in New York which represents more than 600 American music publishers) laid a criminal complaint against the International Lyric Server (http://www.lyrics.ch/) located in Basel, Switzerland and a Swiss District Attorney ordered the seizure of ILS's computers as evidence in the upcoming trial. See http://www2.echo.lu/legal/en/news/9901/chapter10.html#3 .

[9]The German branch of the International Federation of the Phonographic Industry made this argument in their statement on MP3 piracy. See Note 4, *supra.*

[10]The Canadian Department of Foreign Affairs and Trade implicitly recognized this when it noted in a recent discussion paper that "although digital networks increase possibilities of infringement of rights, it can also give rights holders an added ability to control access to and distribution of their works. Thus, solutions to the challenges will in some cases lie with the rightholders themselves." Canada. Department of Foreign Affairs and International Trade. *Canadian Trade Policy in Electronic Commerce*, April 1999, p. 6.

[11] *R. v. Dyment* (1988), 45 C.C.C. (3d) 244 at 254.

[12]*Hunter v. Southam*, [1984] 11 D.L.R. (4th) 641.

[13]Paragraph 39, *Magna Carta*, 1215.

[14]Under s. 487 of the *Criminal Code*, for example, a police officer may only search a building, receptacle or place once a justice is satisfied that there are reasonable grounds to believe an offence has occurred there or the building, receptacle or place contains evidence pertaining to an offence, and a warrant has been issued.

[15]Standing Committee on Human Rights and the Status of Persons with Disabilities, 35th Parliament, *Evidence*, Meeting No. 15, pp. 12-13

[16]*R. v. Wong* (1990), 1 C.R. (4th) 1, 60 C.C.C. (3d) 460 (S.C.C.), reversing (1987), 56 C.R. (3d) 352, 34 C.C.C. (3d) 51 (Ont. Prov. Ct.).

[17]Whether an individual has a reasonable expectation of privacy is decided upon the totality of the circumstances, which may include: presence at the time of the search; possession or control of the property or place searched; ownership of the property or place; historical use of the property or place; existence of a subjective expectation of privacy; and existence of a subjective expectation of privacy. *R. v. Edwards* (1996), 26 O.R. 736, 192 N.R. 81 (S.C.C.)

[18]The sole exception to this rule is when online communications are encrypted.

[19]Data matching software can search through separate data bases and collate all information related to a particular individual. Data mining software uses artificial intelligence to search separate pools of data and identify previously unsuspected trends of behaviour.

[20](1993), 84 C.C.C. (3d) 203.

[21]*Ibid,* at 213.

[22]Federal: *Privacy Act*, R.S.C. 1985, c.P-21; Alberta: *Freedom of Information and Protection of Privacy Act*, S.A. 1994 c. F-18.5; B.C.: *Freedom of Information and Protection of Privacy Act*, R.S.B.C.1996, c.165, *Privacy Act*, R.S.B.C. 1996, c.373; Manitoba: *Freedom of Information Act*, S.M. 1985-86, c.6 (C.C.S.M.c.F-175), Privacy Act, S.M. 1970, c.74 (C.C.S.M. c.P-125); New Brunswick: *Right to Information Act*, S.N.B. 1978, c. R-10.3, *Right to Information Amendment Act*, S.N.B. 1995, c.51; Nova Scotia: *Freedom of Information and Protection of Privacy Act*, R.S.N.S. 1993, c.5; Ontario: *Freedom of Information and Protection of Privacy Act*, R.S.O. 1990, c.F-31; Quebec: *An Act Respecting Access to Documents Held By Public Bodies and the Protection of Personal Information*, S.Q. 1982 (L.R.Q. c.A-2.1); Saskatchewan: *Freedom of Information and Protection of Privacy Act*, S.S. 1990-91, c.F-22.01, *Privacy Act*, S.S. 1990-91, c.P-24.

[23] *An Act Respecting the Protection of Personal Information in the Private Sector*, S.Q. 1993, c.17

[24]See Colin Bennett, "Rules of the Road and Level Playing Fields: The Politics of Data Protection in Canada's Private Sector", *International Review of Administrative Sciences*, Vol. 62 (December 1996), p. 486.

[25]CAN/CSA-Q830-96.

[26]CAN/CSA, *supra*

[27]*Ibid*, p. 9.

[28]S. 4, *Personal Information Protection and Electronic Documents Act,* 1st Session, 36th Parliament, 46-47 Elizabeth II, 1997-98, as amended.

[29]Industry Canada/Justice Canada, *The Protection of Personal Information: Building Canada's Information Economy and Society* (January 1998)., pp. 2-3

[30]*Ibid*, p. 7.

[31]See the submissions made by the Public Interest Advocacy Centre and the Consumers Association of Canada to the Standing Committee on Industry, 36th Parliament, *Evidence*, meeting no. 81, December 8, 1998.

[32]Section 5(2), *Personal Information Protection and Electronic Documents Act, supra.*
[33]Standing Committee on Industry, 36th Parliament, *Evidence*, meeting no. 9, March 16, 1999.
[34]*Final Act Embodying the Results of the Uruguay Round of Multilateral Negotiations*, done at Marrakech, Morocco, April 15, 1994, Annex 1C: Agreement on Trade-Related Aspects of Intellectual Property Rights, April 15, 1994.
[35]*Diplomatic Conference on Certain Copyright and Neighbouring Rights Questions*, Geneva, December 2 to 20, 1996.
[36]*Ibid.*
[37]The three key issues identified included: whether electronic exchanges over the Internet are goods under the GATT or services under the GATS; how to ensure intellectual property rights are enforced on the Net; and how to resolve conflict of laws and disputes over legal jurisdiction. See Canada. Department of Foreign Affairs and International Trade. "Canadian Trade Policy in Electronic Commerce", supra, p. 7.
[38]*Ibid.*
[39]*The Work Programme on Electronic Commerce: Background Note by the Secretariat*, WTO Documents IP/C/W/128.
[40] European Commission. Green Paper. *Combatting Counterfeiting and Piracy in the Single Market.* No. COM/98/0569/FIN, December 1998.
[41]For a detailed analysis of the impact of the printing press on intellectual property laws, see Shelley Wright, "Academic Freedom and Intellectual Property", paper presented to the *Academic Freedom: History of a Defining Idea Conference*, University of Saskatchewan, September 19-22, 1996.
[42]Section 3(1) of the *Copyright Act* provides that the holder of the copyright has the exclusive right to reproduce, produce, perform, publish, translate, adapt, record or film, communicate, exhibit, sell, rent, or distribute the protected works. R.S.C. 1985, c. C-42, as amended.
[43]Legal Advisory Board. *Report on the Communication from the Commission: Follow-up to the Green Paper on Copyright and Related Rights in the Information Society*, No. A4-297/97.
[44]European Commission. *Green Paper on Copyright and Related Rights in the Information Society*, No. COM(95)0382 - C4-0354/95, pp. 51-52.
[45]Legal Advisory Board, *supra.*
[46]*Ibid.*
[47]Section 6 of the *Copyright Act* provides that copyright subsists for the life of the author plus the end of the year in which the author dies plus fifty years.
[48]Section 29 of the *Copyright Act* permits copying when it constitutes fair dealing with the work. Fair dealing occurs when the conduct is fair and equitable and has no material adverse commercial impact of the holder of the copyright. Specifically, fair dealing for the purpose of private study does not infringe copyright.
[49]Legal Advisory Board, *supra.*
[50]Canada, *Privacy: Where Do We Draw the Line*, Report of the House of Common Standing Committee on Human Rights and the Status of Persons with Disabilities (Ottawa: April 1997)
[51]Wright, *supra.*

Chapter 11

Canadian Telelearning Experiences

LAURA WINER
University of Montreal

1. INTRODUCTION

The term *telelearning* has not yet reached a stable definition. But given its recent emergence and rapid, on-going, evolution this is not surprising. Some use the term to refer to any use of computers for learning, thereby including "traditional" computer-assisted learning applications, and dedicated videoconferencing facilities. For the purposes of this paper, however, we will limit its definition to telecommunications-based learning, including the use of multimedia. Telelearning commonly refers to situations in which learners and instructors are separated from each other by time and/or distance; the core feature is that the learners are using networked computing to access any of a number of informational, educational and computational resources. What is allowing telelearning to grow beyond the narrower limits of dedicated facilities and time and space constraints is the ongoing development of the Information Highway.

The "Information Highway" refers to the cluster of technologies that are shaping the evolution of our society in general, and by extension, instructional interventions. The term Information Highway reflects the convergence of previously separate communications technologies (such as the telephone and television) and computing systems into a single global network of networks (Industry Canada, 1995). The Information Highway thus includes both the Internet (the worldwide network of computer networks) and all other distributed, interactive technologies; i.e. audioconferencing, compressed videoconferencing and computer networked learning. The burgeoning Information Highway is playing a key role in the shift towards increased use of technologies in education and the move to "lifelong learning" rather than age-defined periods of study and training. The Information Highway is in the unusual position of being at the same time a contributing cause to the shift, an effect of the shift, and a potential solution to some of the needs and problems that the shift creates.

By making use of the Information Highway, telelearning encompasses "traditional" distance learning (based on the model of extending the classroom) and distributed learning (a model where learning resources are provided "just in time, and just for me", (Wagner, [1999]). The focus of telelearning thus goes beyond simply extending the classroom via technology to include the needs of individuals looking for immediate access to information, performance support tools, and educational opportunities. While the most important technologies in terms of the distance aspect are video and audio conferencing technologies, the inclusion of the concept of distributed learning brings interactive connectivity, linking learners and resources, to the fore. Different industrial players are involved with the different focuses: while conferencing technologies are most significantly affected by the cable television and telephone industries, technologies required for distributed applications are most heavily affected by developments in the hardware, software and networking industries. Traditional distance education tends to use audio, text, and video separately, whereas in distributed approaches, audio, video and data-transmission converge in an integrated form. Telelearning is the umbrella term that incorporates both these approaches to non-classroom-based education and training.

It is also important to consider the kind of learning referred to in different approaches. Telelearning is not limited to traditional discipline-oriented degree programs. Rather, the whole range of categories of learning that our information-based society requires and encourages may be addressed within a telelearning focus: learning required by an organisation of its employees; learning opportunities offered by an organisation, either internally to its employees or as a service to an external public; education/career focused learning initiated by the individual within a formal structure; self-improvement, recreational learning, and family based learning, also initiated by the learner(s) but often in an informal structure; and community-based learning (Masie, 1999).

Telelearning is blurring many traditional distinctions: traditional learning vs. distance learning; students vs. workers, home vs. work; consumers vs. producers of information; formal vs. informal learning; training vs.

education. At the same time, it may serve to increase other distinctions, most notably, the haves vs. have-nots. And the "haves" are now not limited to money or material possessions, but expanded to include those intangibles that money enables: access to the Information Highway, to the new knowledge economy and to the skills and knowledge required to succeed.

In order to explore the ways in which telelearning is affecting Canadians' educational experiences, and to lay the groundwork for a considered reflection on issues that should be considered in looking at the future of telelearning, we will begin with a brief discussion of current applications in Canada. This is not an exhaustive survey; but simply an overview of the different ways in which telelearning technologies and approaches are influencing learning experiences for both children and adults, at school, at work, and at home. These influences are all—at least in some ways—positive ones, but they do nonetheless raise issues that must be given careful consideration when charting a course for future policies, decisions and actions.

2. APPLICATIONS

Developments in instructional interventions respond to economic, social and research developments. Flexibility, cultural issues and the technology push are now important emerging themes (Winer, Rushby & Vázquez-Abad, 1999). There is movement towards more flexibility in terms of time (both scheduling and pacing), content, entry requirements, instructional approaches and resources, and course delivery and logistics. With increased globalisation has come an increased sensitivity to cultural issues at all levels—local, national, global, and corporate. And the increasing use of technology as a tool in business is effectively pushing for more use of technology in both education and training; in the former to prepare students to be full participants in the technology-rich world of work that awaits them, in the latter to rationalise already acquired resources and increasing demands on time (Winer, Rushby & Vázquez-Abad, 1999).

Telelearning is being used throughout the range of learning situations from formal to non-formal learning to informal. Formal learning refers to systematic, hierarchical learning that takes place within educational institutions and is planned for a specific target population; non-formal learning is also organised with specific learning objectives and target population, but is outside the degree-granting system; informal learning refers to the acquisition of skills, knowledge and/or experience by an individual outside an organised program of study.

Telelearning both supports and is supported by communities' desire to ensure access. Opportunities and impacts on learning can be catalysts to action; frequently people are driven to acquire the technology because they want their children to have access in school. The resources thus acquired and created are themselves often the springboard for involving adults in learning activities. In one case study of five community learning networks (OLT, 1998), researchers found a broad range of technology being used for learning, including formal, non-formal and informal. These included public and private computer networks, the Internet, computer conferencing, videoconferencing, computer-based courseware, and computer-based self-paced instruction. Learning and networking technologies clearly overlap; that which helps create and sustain a community can help create and sustain a learning community.

2.1 The Market

While commonly people say that "everyone" has access to the Internet, it is important to put that "everyone" in perspective. In 1997, approximately 57 million people worldwide had Internet access (representing less than 2% of the world's population), and 14 million of those only had e-mail access (Doyle, 1997). In Canada, while one third of Canadian homes had computers, only 13% used modems to connect to the Internet (Statistics Canada, 1997b). This is, however, almost double the number from 1996 (7.4%), and this rate of growth can be expected to continue or even increase in Canada.

The education market, specifically K-12 and postsecondary, spends significant amounts of money on technology, even if it represents a relatively small percentage of the total monies spent. In 1997-98 in the United States, expenditures were U.S. $5.2 billion (Quality Education Data, cited by Bort, Julie [1998] Schools' learning curve boosts networking market," Electronic Business, 24[1] in Curry et al., 1998). The amount of money spent in Canada on technology is significantly lower; however, Canada is a leader among the G8 countries in connecting its schools to the Internet. SchoolNet, sponsored by Industry Canada, has linked all Canadian schools and libraries to the Internet (www.schoolnet.ca). The fact that all primary, secondary, and post-secondary institutions in Canada have at least some limited access to the Internet means that energies spent on improving the quality of materials available on the Internet will theoretically have benefits for all.

Many postsecondary institutions have embraced the Internet as a way to deliver or enhance courses, for both on-campus and distance offerings. TeleEducationNB, for example, lists dozens of sites with links to innumerable courses and resources for producers and users of on-line educational services (http://teleeducation.nb.ca/home/). The World Lecture Hall presents links through over 70 different topics to faculty worldwide who are using the Web to deliver course materials (http://www.utexas.edu/world/lecture/). WebCT, based at UBC, has a list of over 679 institutions in 37 countries using WebCT to support teaching and learning activities (http://www.webct.com/). The majority of the WebCT users are in the U.S., but they are by no means limited to the English speaking world, with such diverse regions as Japan, Slovenia and Sweden represented.

The education market is not, however, the only one involved in using new technologies as a learning tool. Indeed, many sectors of the economy are more technologically sophisticated (for example, health services, banking, environmental industries, printing and publishing, electrical and electronic products industries, software and computer professionals, agriculture and construction) and are consequently more amenable to incorporating new multimedia learning materials (NMLM) that may or may not be delivered in a telelearning context.

A number of factors are influencing the growth in use of learning technologies. Based on extensive literature reviews, case studies and expert input, Ekos & Green (1999) identified a number of factors influencing growth in the use of learning technologies in general. Those presented below contribute both to increased interest in and application of telelearning.

- Decreasing delivery costs: both hardware and access costs continue to decline.
- Decreasing development costs: improved development tools and increased development experience mean that the costs of creating instructional materials are declining.
- Increasing availability of off-the-shelf products: while many markets are still too specialised or too unstable to benefit from pre-packaged materials, there is a growing bank of technology-supported learning and training materials that can be used as is or easily adapted for use.
- Increasing expertise in educators and trainers in development, delivery and use of technology: teachers and trainers themselves are becoming more familiar with new technologies, thereby reducing the costs of upgrading their skills to allow them to work with new materials.
- Growing awareness in the population (employers, employees, students, teachers, parents) about the effectiveness of learning technologies: This leads to a push for access to more learning technology by the learners, as well as increased receptivity to, and tolerance for, growing pains when the technology is introduced.
- Emergence of parallel suppliers to traditional sources of education/training: As more people have access to the Internet, whether

from work, home or community centres, traditional local sources no longer have a captive market, and people are no longer limited to what is available locally.

- Growing awareness of the importance of individual learning styles: The ability to provide instruction for people to learn at their own pace, in their own way, and in privacy, responds to increasing concerns about how best to tailor instruction to the diverse needs of our population.
- Chicken and egg—more technology needs more training: As technology becomes more prevalent in the home and workplace, more training will be needed. It seems only logical to provide this training using the technology itself.
- Productivity push leads to more training, leads to more technology: Parallel to the chicken and egg factor is the push to improve productivity—both in industry and education. Technology is seen as the way to improve productivity; in turn the increased use of technology increases the need for training.

2.2 Training and Adult Education

The training market is growing. It is forecast that workers will soon have to spend at least 20% of their time in learning (Curry et al., 1998); a significant percentage of this may be on their own time and initiative, even though it will be directly related to their work. Currently, although corporate training expenditures are increasing, much of this training is still occurring in traditional classroom-based contexts (ASTD, 1998; Winer, Rushby & Vázquez-Abad, 1999). Many companies are, however, expressing interest in incorporating more use of technology into their training repertoires.

According to Statistics Canada, half of adult education is employer-supported (Statistics Canada, 1997a). Although not as large as the K-12/postsecondary education market, the training/adult education market is the one that is seen as the main driver for growth. In Canada, the 40-64 age group will grow from 8.7 million to 10.2 million between 1996 and 2001 (Statistics Canada, 1998). Not surprisingly, the average age of university undergraduates is increasing: approximately 40% of all people pursuing formal degree programs in the U.S. are over 40 (Curry *et al.*, 1998).

Other important influences include the shift in work patterns to higher levels and longer periods of unemployment as well as the fact that part-time work is now a major feature of Canada's economy (Industry Canada, 1997). These raise the question of what the implications are for access to training, updating skills, etc. On-going skills and knowledge development, or lifelong learning, will be an economic necessity as well as something perceived as being for the general social and individual good. If an individual is kept out of the training loop, his or her skills will be not be upgraded and getting back or staying in the employment loop may become increasingly difficult. This importance of lifelong learning will thus most likely fuel the demand for more options in participating in learning activities, thereby stimulating the demand for more telelearning options, with their inherent flexibility to adapt to the myriad needs and constraints of adult learners.

As Ekos and Green (1999) point out, job-related training can be delivered using learning technologies at schools and at home to the home-based self-employed, another important and growing sector of the workforce. In addition, employers' increasing reliance on flexible work practices requires continuous learning of new job skills that in turn requires courses to be developed and delivered in short periods of time. Downsizing, outsourcing, non-standard positions (part-time, job sharing, casual, etc.) are all aspects of the current employment climate that limit access for many to workplace-based training.

Telelearning is one way that individuals can have access to the training they need outside of a workplace setting. The challenge is, of course, in connecting individual people to the learning resources they need. One example of how the Information Highway can help is the LearnTec project, in which an online compendium of resources and practices was developed to provide access to CD-ROM and Internet sites that provide or complement vocational and trades training (http://www.cotr.bc.ca/online/learntec.htm). The richness of this resource can be seen in that while its base is in British Columbia, service providers are in Canada, the U.S., Great Britain, Australia and New Zealand.

Ekos and Green (1999) also examined the perceived benefits for employers for all training activities involving new media. The highest rated were:

- Improved employee satisfaction with training
- Increased training effectiveness
- Greater flexibility
- Self-pace
- Uniformity of training

Telelearning capitalises on many of these benefits. The potential for this kind of training to attract and retain qualified employees was perceived as an additional benefit.

It is interesting to examine what benefits were seen as less important by employers, given the usual stress placed on the bottom line. For employers, the lowest rated benefits were:

- Reduced training time
- Reduced training costs
- Access to training in remote sites or off-site

2.3 Education

2.3.1 K-12

There are many initiatives addressing the K-12 market. The one that affects the highest number of Canadians is SchoolNet (www.schoolnet.ca), which has succeeded in connecting all of Canada's K-12 schools and libraries to the Internet. SchoolNet began in 1993, and operates in partnership with provincial and territorial ministries of education, library authorities, professional associations, and the private sector. The SchoolNet initiative has contributed significantly to Canada's leading place among G8 countries in terms of school and library connectivity. The provinces also have established their own networks; see for example, the Manitoba Educational Research and Learning Information Networks (MERLIN), established to coordinate the use of technology for learning across the province of Manitoba (http://www.merlin.mb.ca/) and Cyberscol in Québec which performs a similar function (http://www.cyberscol.qc.ca).

Schools in the Canadian K-12 sector are using telelearning in many different ways. One of the most common uses is an extension of the pen-pal model, with students in one school corresponding with students elsewhere in their province, across Canada, or around the world. However, with the sophisticated technology readily available, images (still and video) and sound often accompany textual exchanges. Students use the Web as a complement to their school or local library; they also participate with experts or peers in projects designed to link interested participants in genuine enquiry (See the Biosphere in Montreal for one example; http://biosphere.ec.gc.ca/cea/default_a.html.)

2.3.2 Post-secondary

Telelearning is having a significant impact on post-secondary institutions in Canada. Institutions historically involved in distance education such as Athabasca University in Alberta (http://www.athabascau.ca/), Télé-université in Québec (http://www.teluq.uquebec.ca) and the Open Learning Agency in British Columbia (http://www.ola.bc.ca/) are integrating the possibilities of new technologies into their established repertoire of courses. Telelearning is also facilitating the entry of historically campus-based institutions across Canada into a non-campus-based mode. This push to offer non-campus courses is a result of many factors (for example, shifting demographics, interest in lifelong learning), some of which were discussed above. There is an important and growing sector of the population interested in pursuing university-level studies but who for a number of reasons are unable or unwilling to follow traditional campus degree programmes. With the globalisation of education, students anywhere in Canada with a computer and a modem have access to courses offered by institutions around the world. Canadian universities no longer have a captive market for students who want the flexibility that telelearning offers. At the same time, Canadian universities can themselves expand beyond their local boundaries to attract students across the country and around the world.

Telelearning is not limited, however, to off-campus courses. There is a growing trend towards a mixed mode of learning, integrating face-to-face campus courses with telelearning components (Bracewell et al., 1998). Most, if not all, Canadian universities are encouraging professors to make use of telelearning in support of their on-campus teaching. This may seem like an oxymoron, but there is growing interest on the part of students, professors, and administrators in integrating synchronous and asynchronous computing resources for content presentation, access to searchable information, information exchange, guidance, practice and feedback, discussions, and simulations (Gandell, Weston, Finkelstein & Winer, in preparation).

Another factor supporting the development of telelearning applications is the availability of shells that facilitate the development of courses or whole programs. Such shells minimise the need for programming knowledge, and allow professors to exert a similar level of control over their course development as with a less technologically intense one. These shells have come either from academia, as in the case of WebCT from the University of British Columbia (http://www.webct.com/webct/), or from private industry as in the case of Learning Space from Lotus (http://www.lotus.com/home.nsf/rightframe/1education).

2.3.3 Professional educators

A corollary to developments in education is the necessity for on-going professional development. Numerous listservs and discussion groups provide forums for teachers and administrators to share and discuss their experiences and build on-line communities that support on-line learning. One example is the Education Network of Ontario (ENO) (www.enoreo.on.ca), an independent, non-profit joint project of the Ministry of Education and Training, the Ministry of Economic Development and Trade and Tourism, and the Ontario Teachers' Federation, designed to enhance school-based educator professionalism through electronic networking. Another example is the BC Educational Technology Users' Groups (http://www.ctt.bc.ca/edtech/), informal groups of provincial post-secondary educators, administrators and technical personnel who meet, face to face or on-line, to deepen their mastery of learning technologies and promote the incorporation of technologies into teaching activities. Both SchoolNet and Cyberscol have resources and forums for teachers to share professional development experiences.

2.4 Others

Another growth area is that of Community Learning Networks, community-based and controlled initiatives whose main purpose is to further the development and enhance the lives of community members, through formal, informal and non-formal educational activities. Examples of these are found across the country in small and large geographic communities, serving a broad range of people and interests.

There are many instances of ICT-enabled (Information and Communications Technology) community development efforts. One is found in northern Saskatchewan for a population that would not normally have access to this kind of learning experience. The "Community Toolbox Model of Economic Development", a course normally delivered in a classroom setting, is being modified and adapted for Internet delivery (www.idrc.ca/acacia/nordicit/title.htm). This course integrates visiting mentors, peers and online instructors using Internet-based video-and audioconferencing workshops.

Contact North/Contact Nord in Ontario is a not-for-profit corporation whose formal mandate is improving access to formal and informal education and training opportunities for residents of northern Ontario (www.cnorth.educ.on.ca).

SAVIE's (Société pour l'apprentissage à vie http://savie.teluq.uquebec.ca) mission is to bring together provincial groups that are committed to using information technology for training, as well as to encourage and support community innovation and learning. Participants work on common training problems and challenges, and jointly develop training tools and technological infrastructure.

Connect NB Branché (New Brunswick; http://cnbb.unb.ca/english/) opens up public-sector institutions such as schools for use as Community Access Program (CAP) sites. Sites are advertised as learning centres; their goal is to provide a range of educational services to the local community, including Internet access, for a modest fee.

The mandate of the Télécommunauté Insulaire Francophone (Prince Edward Island; http://www.teleco.org/640-480/anglais/index.html) is to provide access for adult francophones in PEI to educational services in French. Educational services targeted include postsecondary, professional, community and life skills. The Telecommunity currently performs an information dissemination function; it will eventually provide links to telelearning courses through institutions such as University of Moncton.

In the far north, the Igalaaq, Rankin Inlet Community Access Centre (Northwest Territories; http://www.arctic.ca/LUS/CAC.html) provides community access to computers and the Internet. This network facilitates distance education with McGill University in Montreal.

A last example is the On Line, In Business, At Home (OLIBAH) project in British Columbia that was designed to assist adult learners with various types of disabilities in obtaining access to education, career and information technology training (www.openschool.bc.ca/olibah/kit).

2.5 Home

In 1997, 36% of Canadian households reported having a computer but only 13% had a modem and used it to connect to the Internet (Green, 1997). Not surprisingly, the likelihood of both having a computer and using it for the Internet correlate positively with income level.

However, there are other important influences, with geographic location also being very important. For example, 66% of 25-44 year olds in urban areas, as opposed to 54% of those in rural areas, have access. And even though almost all Canadians have access to phone lines, about 3% of those phone lines still require conversion to single-party lines that support touch-tone, and therefore computers. This is especially true in northern and remote communities.

Age and education level are also important factors in individuals' ability to use computers. In 1994, while 84% of 15-19 year olds reported being able to do "something" on a computer; only 36% of 55-64 and 10% of those over 65 reported the same level of familiarity.

The more educated one is, the more likely one is to be computer literate. Of Canadians with an MA or PhD, 84% can use a computer; however, 50% of those with high school only can, and 31% with less than high school can.

Among those who are currently employed, 67% can use computers; among unemployed only 50%.

And finally, the use of online services is more common among men than women: almost 60% of users of online services are men, and men account for 63% of total usage.

3. ISSUES

3.1 Canada's Role

Canada has long been recognised as a world leader in telecommunications; and to some extent this has carried over into telelearning. Canadians pioneered developments in online design and delivery platforms. (See for example Virtual-U www.vlei.com and WebCT www.webct.com.) However, Canada is not leading the way in offering specific courses or programs; many of the links from Canadian sites such as TeleEducationNB, WebCT, Virtual U, and others lead to U.S. or other international institutions.

Because people can now enrol at what are perceived as more prestigious U.S. post-secondary institutions, it is important for Canadian institutions to achieve greater visibility and credibility on the world stage. Otherwise we risk a run-off of students to non-Canadian based institutions, and subsequent issues of intangible cultural differences for school-based curricula, as well as the relevance and fit of technical training and norms. For example, safety and environmental standards often differ significantly between countries—does the availability of "off-the-shelf" training present a risk that Canadian standards may be undermined? Within the educational system, questions arise as to who will be writing economic policy and history courses. Will they respect what our society determines to be the appropriate perspective on international events?

3.2 Traditional teaching

Telelearning, by facilitating and enabling new modes of communication, places additional pressures on the traditional school system. Unesco has identified Information and Communication Technologies as having the most potential for significant impact for conventional education as a medium for interaction between teachers and learners, among learners, and among teachers (Bracewell et al., 1998). Groups such as the Canadian Teacher Federation have seen the influx of technology as an opportunity to discuss fundamental issues facing the educational system, and are using the technology issues as a catalyst for discussion (www.cft-fce.ca/e/what/restech/ART&PAP2.htm). At the same time, shifts in workplace skills and modes (for example, more focus on collaborative group work, especially computer supported collaborative work) will have an impact on teaching as pressures from the "real world" are felt inside the schools. Parallel to these economic and social developments, the current state of the art in learning theories presents a shift to constructivist approaches and away from the knowledge transmission model of teaching. Computers are often presented as offering a way to bring the world to school, but it is important to remember that when the world comes to school, one has a window out as well as a window in. This openness means that students, teachers and administrators are faced with new skills, knowledge, and ethical questions. Along with the move towards a vision of lifelong learning must be a re-examination of the traditional goals of the public school system. If there is a general recognition that graduation (whether high school or university) is not the end of an individual's life as an intentional learner, then what are the learning skills and attitudes students need to develop within these institutions to serve them later?

All these issues must be seen in the context of real learning outcomes, that is, what we must pay attention to is what individuals are actually *learning* rather than what is being taught. This leads to the question of what we evaluate, and how. We have traditionally focused on evaluating the *product* of learning activities—how do we expand our repertoire of tools to adequately assess the *process* of learning?

3.3 Vanishing borders

The telelearning environment escapes traditional boundaries of space, time, pace, mode, level and role. It leads to an introduction of market forces, with universities and other institutions (no longer local) competing for students. At the same time, the market for K-12 educational resources is expanding. All this means that traditionally circumscribed areas are suddenly open for bi-directional activity.

3.3.1 Geographic

As soon as learners are on-line, they have access to resources that are geographically dispersed. The learner is now limited more by language of instruction than by geographic location. This means that learners, especially minority language speakers, have more resources available to them. For example, anglophones in Québec, francophones outside Québec, and non-French or English speakers across Canada have many more options open to them. Remote communities no longer have to rely only on what is physically accessible, but can participate in on-line learning activities that fulfil their learning goals regardless of the location of the site that offers them. However, more subtle issues, such as variations in curriculum, approaches and values, may be difficult to see, although important over the long-term.

3.3.2 Social

We are used to having clear boundaries between teacher and student, trainer and trainee. However, in many cases these boundaries are now blurry and in flux. Participants in telelearning can take on different roles, depending on the context and purpose of their interaction. There are now real possibilities for more collaboration and active participation, and in previously unimaginable contexts.

The blurring of geographic borders mentioned above may have an impact on social cohesion. The Canadian Radio-television and Telecommunications Commission (CRTC) has announced that it will not attempt to regulate the Internet (CRTC, 1999). This means that Canadian content will not be assured as it is on television and radio. This raises a number of questions. Who will be educating our children? Our employees? What cultural models of acceptable practice will be instilled? What will the knowledge base be?

The CRTC announcement is not the end of decisions for our society. The broad question remains of which stakeholders will define, regulate and develop what the Information Highway will be used for. How will public needs be balanced against the "bottom line"?

3.3.3 Organisational

With the blurring of distinctions between education and training, the roles of the players are not always clear. Which are educational institutions, and which are profit-making private industry? Should certain roles, such as K-12 curriculum provider, be reserved for non-profit players? Who are the stakeholders?

Degree granting institutions such as colleges and universities currently have residency requirements. While in the past, this has clearly referred to physical residency, what will happen to that requirement as telelearning expands into campus-based courses?

3.4 Lifelong learning

The concept of lifelong learning recognises that individuals learn intentionally and in an organised manner over the course of their lifetime, in a variety of contexts. Traditional educational institutions may be the means through which lifelong learning happens, but not necessarily the only one.

Lifelong learning means that a whole new pool of learners is created. These include post baccalaureate adult learners trying to stay up-to-date in their chosen discipline, as well as adults trying to acquire new skills and competencies in view of a career change (whether voluntary or forced).

3.5 Polarisation

Polarisation can happen in many areas: age, culture, formal education, level of literacy, degree of marginalisation within the mainstream culture, SES, regional differences, linguistic differences, part-time vs. full-time workers. Access, both widespread and equitable, is the key to reducing the polarisation of our society. This requires a serious consideration of what kind of access is needed. What bandwidth is really needed for valid educational uses as opposed to what is being pushed for entertainment and commercial applications? It is also important to distinguish quantity of information and programs from their quality, and to judge overall access by the value of the information that you have access to.

4. QUESTIONS

The above issues lead inevitably to many more questions than answers. These are the questions that we as educators, trainers, researchers, policy makers and citizens will have to address as the importance and pervasiveness of telelearning continues to grow. They are listed below in the hopes that they will inform our thinking and our actions.

4.1 Access

- How will physical access be provided to all Canadians, in terms of access to equipment and servers?
- How will functional access be assured, i.e., ensuring that potential users have the necessary knowledge, skills and awareness?
- How will equitable access be addressed? How welcoming is the entry point for minority learners (SES, cultural background, disabilities, etc.)? How age friendly is the access (young and old)? How gender neutral?
- What comprises informed access? What safeguards will there be for privacy? How is learning interaction being tracked, monitored, etc.?

4.2 Cost

- Who will be paying for the technology (initial purchase, ongoing maintenance and upgrades, hidden costs)?
- Who will cover related costs (e.g. childcare)?
- What are the best uses of resources for different targeted populations?

- How will cost-effectiveness be evaluated? If the learning activity is ongoing as opposed to a fixed length of time, how is it accounted for?
- Who will pay for training—the employer or the employee?

4.3 Time

- When is training done—on whose time, the employer's or employees'?
- What are the implications of access anywhere anytime for non-work time? What will the social fall-out be in terms of family time?
- Where will the time come from for keeping up with ever-changing technology?

4.4 Quality and equality of learning

- What are the impacts of telelearning developments on the quality of learning activities and equality for the population as a whole?
- How transferable is the technology and skills/expertise?
- Who ensures quality control?
- Innovation and change:
- What is the real opportunity and for whom?
- Is it sustainable?
- What are the opportunity costs?

5. CONCLUSION

It may be initially surprising how few of the issues raised are strictly educational concerns. But the fundamental goal of education and training is to support individuals—and by extension groups, communities, organisations and our society—to survive in, adapt to, and ultimately influence for the better, our world, at local, national, and global levels. Education is clearly a societal system that cannot be examined in isolation or in only one discipline. Research in psychology will guide us as to micro level implementation strategies for design and use. Developments in computer science and telecommunications will provide us with new tools and means to implement our chosen strategies. As a society, however, we must examine the issues affected by technological change, and not simply go unquestioningly where the technology beckons.

Chapter 12

The Productivity Paradox

ICTs, Knowledge and the Labour Market

NICO STEHR
Max-Planck-Institut für Meteorologie

1. INTRODUCTION

When it comes to the economic affluence of a nation and the ability of a country's economy to improve the standard of living of its citizens and compete internationally, social scientists are in an unusual agreement that productivity "in the long run is almost everything" (Krugman, 1994:13). Manuel Castell (1996:80) throughout his extensive study of modern society as a network society seconds this observation and concludes that "productivity is the source of the wealth of nations".[1] Not only shifts in standards of living follow from changes in productivity performance. Less immediately related non-economic transformations in response to unequal national productivity gains occur, including major changes in the balance of global power relations. In this light, Krugman (1994:17) comments somewhat despairingly, the slowdown of "American productivity growth since the early 1970s becomes the most important single fact about our economy."

Global competition represents the linkage path that combines information and communication technology (ICT), organizational change and productivity growth. Productivity gains are stratified. Productivity improvements or the lack thereof do not occur in a linear fashion throughout the economy nor among all divisions within a firm or segments of the work force. In the advanced economies the challenge is no longer "to make manual work productive… the central challenge is to make knowledge workers productive" (Drucker, 1999:141). And the tools modern knowledge workers increasingly deploy are information and communication technologies.

A similar equally widespread agreement among economists, policy makers, sociologists and educators extends to the notion that "structural changes" in the modern economy transform the labor market and generate demands for specific skills. Changes associated with the emergence of the knowledge-intensive economy are seen to engender an increased need for skilled workers (e.g. Baldwin and Gellatly, 1998: v) or a growing presence of knowledge workers in the economy. [2]

Three decades ago, Daniel Bell and others began to explicate their idea that industrial society is giving way to a post-industrial society. One of their central tenets was to refer to the centrality of theoretical knowledge as constitutive for post-industrial society. According to Bell (1968:158), the crucial political questions post-industrial society will face "deal with education, talent, and science policy. The rapid expansion of a professional and technical class, and the increased dependence of the society on scientific manpower, suggest a new and unique dimension in social affairs: i.e., that the economic growth ratc of a postindustrial society will be less dependent on money than on 'human capital'."

In the discussion and controversies surrounding the productivity paradox, these and related observations about the decline in the economic importance of the forces of production that shaped industrial society are rarely moved toward the center of the analysis or linked to observations about the nature of the productivity puzzle even though they now form the premise of many general analyses of the modern economy as a knowledge-based economy.

I will try to link the analysis of the productivity paradox in this paper to the emergence of knowledge as a source of economic growth and changes. I will argue that the productivity paradox can be better understood if one recognizes three empirical facts. First, highly skilled labor appears on the scene before information technology. Second, the increasing importance of highly skilled labor is not a reaction to demand for such labor, but rather there is an autonomous (i.e. societally driven) supply shift. And third, information technology actually helps entrepreneurs and managers to catch up with and reverse the rising labor costs implied by this supply shift. Therefore, the productivity paradox can help us to understand that we are not faced with a technology driven transition from industrial to informational society but rather with a societally driven transition from an industrial to a knowledge society. In this sense, then, we have entered a new modernity.

2. THE PRODUCTIVITY PARADOX

In the last two decades economists in particular have been puzzled and even irritated about the apparent lack of measurable productivity gains in goods producing and services industries in OECD countries in response to or in conjunction with the immense investments in recent years in information and communication technologies. The choice of labelling this phenomenon the "productivity paradox" results from the disjuncture between the immense economic expectations and promises that have been engendered by the "computer age", on the one hand, and the apparent lack of sustainable economic payoffs resulting from the enormous investments by corporations and the state in information and communication technologies, on the other hand.

In 1990 alone, U.S. businesses invested $61 billion in hardware, $18 billion in software and $75 billion in data processing and computer services (U.S. Department of Commerce, 1991). Attewell (1994:24) sums up previous research on the productivity paradox affirming its existence and comments that "no study documents substantial IT effects on productivity." Although conceptual, methodological and data difficulties that extend to the very definition of productivity do exist in the information that are typically utilized in generating these findings, they do not appear to invalidate the results completely.[3] Given the excessive and often repeated claims about the transformative capacity of information technologies, one could even be tempted to ask, why have productivity gains that can be attributed to ITCs not been even more spectacular?

In an intellectual version of the continental drift, the social sciences disciplines are moving farther and farther apart (cp. Luhmann, [1991] 1993:2). Given the socio-cognitive state of contemporary social science, it is not surprising that the growing literature on conditions for the possibility of "productivity growth", let alone the nature and consequences of the "productivity paradox" is not accompanied by a common conception of its meaning and empirical referents. There is no agreement on why productivity varies, let alone on what and how one might account for the essentially contested observation that the growing diffusion of information and communications technologies at work have not improved the productivity of firms as measured by official statistics. For example, is the productivity paradox in the field of ICT a signal that mirrors a more global signal that displays a secular decline in productivity gains? Distinctive research cultures and networks have grown up around the notion of the productivity paradox; networks that do not communicate with each other and pursue their own strategies in examining the issue of the productivity payoff of modern information and communication technologies. These points have been elaborated in the proceedings of a recent OECD conference, The Social Sciences at a Turning Point?, and particularly in the contribution of Langenhove (1999).

The variety of research and accounts of the productivity paradox exemplify the growing and deepening division of labor in social science and its essentially contested nature. For some observers, the productivity paradox does not exist in reality. The productivity puzzle is a measurement construct or indicative of a mismeasurement of outputs that conceals real gains that are made (cp. Quinn, 1996; Diewert and Fox, 1997). But even if the puzzle should exist, the magnitude of the problem is small upon first examination since investments into computers form a relatively minor part of all capital input. For others although the paradox is real its represents but a transitory phase not unlike the productivity lag produced by the transition to technological systems in the past such as the diffusion of electric power. And as is the case for other learning processes, it takes a protracted period of time before the economic benefits show up (cp. David, 1990; Petit and Soete, 1997; Davenport, 1997). Still other observers see the productivity paradox not as a gap that reflects economic realities. Its persistence is rather an indicator of intellectual or theoretical deficits in economic discourse (cp. Jorgenson, 1997).[4] Last but not least, some economists have signaled that the danger has passed and that the productivity paradox disappeared by 1991 (e.g. Brynjolfsson and Hitt, 1996; Sichel, 1999).

Sociologists and scholars from other disciplines who have paid attention to the issue of the productivity paradox have attempted to explain (de-aggregate) the paradox in a great variety of ways. In particular, they have tried to identify various social and organizational mechanisms that undercut or stretch out the potential productivity payoffs from ICTs in firms. Attewell (1994:48) suggests that productivity payoffs from investments in information technology may result from a series of trade-offs within firms at both the individual as well as collective level. For example, the "potential benefits of the technology may be channeled into alternative directions—either doing the original work more efficiently (productivity enhancing) or doing a different kind of activity or the same activity more often." Pinsonneault (1998) has pointed to the differential patterns of association between the usage of information technologies and the nature of managerial work in different firms.

3. EMPLOYING INFORMATION AND COMMUNICATION TECHNOLOGIES

I would like to frame the issue of the productivity paradox in a radically different manner leaving neoclassical conceptions of rational economic conduct and of work behind. Perhaps the most fundamental yet unexamined assumption of the essentially contested accounts of the productivity paradox is the idea that the deployment of information and communication technologies entirely follows the logic of all economic behavior. Investments into tangible and intangible capital is motivated up front and primarily driven by the desire to enhance the profitability, competitiveness, and productivity of firms that decide to embark on such investments.

There is no reason to doubt that such motives play a key role in investment behavior and that expectations of senior managers who invest in information technologies are that robust gains in productivity will be commensurate with the cost of modern ITCs. In the absence of such expectations it makes no sense to speak of a productivity puzzle.

The question therefore becomes whether conventional economic motives and forces are the exclusive or even main reason for the heavy commitment of funds into ICTs. It is not only the possible lack of any visible and robust payoffs in terms of the orthodox motives of economic conduct, or the kinds of everyday frustrations that accompany the use of information technologies (cp. Landauer, 1995) and often speedy obsolescence of skills that generate doubt about the persisting force or viability of such normative dependencies and asymmetries in the relation between the employment of new technologies, the nature of the world of work, production processes and profits.

In a knowledge-based economy other, and as I want to stress, *additional* processes[5] are at work that account for the transformation of the work place and the world of work and therefore for the growth and employment of information and communication technologies even though global and/or local empirical signals as far as the payoff, as measured by conventional economic indicators, are difficult if not impossible to obtain.

If it can be shown that such additional reasons also and increasingly account for the widespread marshalling of ITCs at work, it follows that the productivity paradox assuming that it is real can be evaluated in a very different light and has to say the least, multiple causes.

The main weakness of neoclassical perspectives of work is the failure to attend to the social context of work. Key qualitative features of the world of work such as job satisfaction, social networks, management style and the quality of the products and services, are left out. In fact, "social context and social connections envelop the world of work" (Tilly and Tilly, 1998:13). In knowledge-intensive economies, I would want to maintain the relevant social context of work acquires even greater significance for the competitiveness of a firm, its innovative capacity and profitability. Differences in social context that should not be simply stripped away are of particular importance as one explores differences in the world of work across time. I want to explore the possibility that the productivity paradox constitutes a global signal for a dramatic transformation of the economy generally and the world of work characteristic of that economy in particular.

4. THE LOSS OF AGENCY OR THE WORLD OF WORK IN INDUSTRIAL SOCIETY

The meaning associated with the term "labor" or "work" today is a product of industrial society.[6] The dominant perspective of industrial society sees its technological regime closely linked to mass production systems, intensive productivity gains and the capacity to produce an abundance of goods as well as hierarchical forms of work organization and control. In the end, the (capitalist) logic always contributes to a massive alienation of workers, or as argued more recently, to an extensive de-skilling of the work force (e.g. Braverman, 1974; Gill, 1985). Although the lament about the de-skilling of the individual worker often is associated with the work of Harry Braverman, there are numerous predecessors. The predecessors include Helmut Schelsky. Schelsky (1954:20) asserts: "The closer we approach automation, though without ever fully reaching it, the greater the degree to which work becomes spiritless and stressful and the lesser the extent to which it requires interest in technical matters and skills or, even initiative of any sort." Even earlier, in an essay entitled "The machine, the worker and the engineer," Robert K. Merton (1947:80) refers to numerous social implications of labor-saving technology including the "enforced obsolescence of skills" and the loss of status as well as self-image that accompanies the de-skilling process of workers.

Merton assumes, as does the later de-skilling thesis, that the obsolescence of skills is irreversible. In light of the kind of production technology used, a compensating process presumably is not considered likely. The increasing employment of laborsaving technology produces the enforced obsolescence of skills among the workers. The social and psychological consequences of discarding acquired skills are mainly connected to the demotion of status (including the possible loss of the public identity of the job) and the destruction of the positive self-image of the worker, stemming from the once confident use of those skills. Merton (1947:80) anticipates an "alienation of workers from their job and the importance of wages as the chief symbol of social status are both furthered by the absence of asocial meaning attributable to the task. Increased specialization of production leads inescapably to a greater need for predictability of work behavior and, therefore, for *increased discipline in the workplace*."

Massive power asymmetries in work organizations exist and the struggle for power is a zero-sum game. The ability of management to preserve and exercise domination is assured by virtue of holding on to or monopolizing *knowledge* about the conceptions on which production is based. Knowledge is located in specialized departments only. That is, the successful separation of execution and conception is the key to the control and the persistent degradation of the worker. Employees are mere executioners of tightly prescribed and increasingly fragmented tasks. The new version of the oppression thesis also generalizes about the work place without any credit to the imagination of the worker and specific conditions of work. Any resistance by employees that may be evident is merely in response to the oppressive control exercised by management. The thesis minimizes, consistent with Marx's portrait of the labor process in capitalist society, the active ability of the worker to affect his or her working conditions. Technological developments simply reproduce the domination of capital over labor, often on a more repressive scale contributing to what Merton (1947:80) as indicated called an "enforced obsolescence of skills." The depressing conclusion therefore can only be that work in the modern factory continues to be trivialized and that the "Taylorist philosophy is in many cases still being carried over to the era of microelectronics systems in manufacturing" (Gill, 1985:87).[7]

5. AGENCY OR THE WORLD OF WORK IN KNOWLEDGE SOCIETIES

It is by no means a novel observation that the social organization of work is changing and that the nature of the change has to do with what originally constituted, at least according to Marx and Engels, the condition for the possibility of the division of labor in society: the separation of labor into manual and intellectual labor (cf. Marx and Engels, [1932] 1960:28). Although physical labor and the expenditure of human energy in such tasks will not disappear entirely in knowledge societies, the dominant trend is away from manual labor to intellectual labor, and therefore to a corresponding increase in the role knowledge and learning play in shaping work, the ability to work and continued employment.

More recently, fascination with the constraining features of the social and material conditions which allegedly give rise to persistent hierarchy and isolation of the individual in the workplace, have been somewhat replaced by equally strong convictions about a *new* division of labor, a new technology, and new logics of organizing production that are described as permissive forms of domination (cf. Sabel, 1991:24; Hirst and Zeitlin, 1991; Sabel, 1995). The social distribution of power in the workplace may, as a result, no longer constitute a zero-sum game. These views are linked in turn to the distinction between the declining regime of mass manufacturing ("Fordism")[8] and the growing system of "flexible specialization" in production (cf. Piore and Sabel, 1984). As a result, technology is not seen as an essentially dehumanizing force but as one that enables or at least holds the promise of participation in the affairs of work. This in turn moves the notion of the growing condition for the possibility and importance of "subjectivized conduct of work" (Böhle, 1998:241) into the center of the analysis of the world of work.

The extensive use of technology does displace least-skilled work. For example, in a case study of the impact of computer-aided design on skills in U.S. aircraft and automobile manufacturing industry, Salzman (1989:260) finds that "the technology is relatively effective at...automating the least skilled work (the simplest connections in this case) leaving only the most skilled aspects for the designer." Whether one should really be concerned about such a change is a different matter. It certainly does not represent an instance of de-skilling. Computer-aided design is not the "Trojan Horse of Taylorism" but its opposite, a counter-tendency to a Taylorist organization of work.

In the sphere of work, the profound anxieties about the destructive ways of technology are now replaced by animated discussions about freedoms from control. The vocabulary of intentionality and agency or collaboration and working together, thought to be obsolete, reappears in discussions of work, production and the social organization of work (e.g. Cavestro, 1989). Paradoxically, the technology once feared to have become self-regulating now (de-)regulates itself in the sense of negating regulation.

The distribution and utilization (deployment) of knowledge is not necessarily as one-sided as accounts have stressed the ease with which superiors in modern factories and offices manufacture consent and impose control. Knowledge as a capacity for action is not easily monopolized. Nor is knowledge a one-dimensional and static phenomenon. Knowledge is contingent, contestable, multiple and shifting. Subordinates do not always need to control as much knowledge as their superiors perhaps do in distinct instances or precisely the same forms[9] in order to effectively resist and override controls imposed by management and owners.

If it is indeed the case that modern technology enlarges the capacity to act not only among management, but also among its employees, it would follow that employment regimes that rely on command, control and coercion to ensure performance constitute an increasingly ineffective basis for the coordination of production. Social relations in the workplace could therefore be best shaped by what Max Weber ([1913] 1981) has called *Einverständnishandeln*, or social action that derives from mutual trust. Social relations in the workplace that are based on trust may displace those linked to its opposite, namely distrust. Distrust is, of course, at the heart of the social organization of work that relies on the principles developed at the beginning of the century by Frederick W. Taylor or Henry Ford. Mutual expectations that invoke trust rather than distrust may not only be more typical in organizations that require considerable flexibility, initiative, and autonomy (cf. Heisig and Littek, 1995), but could also be a "rational" response by management to enlarged capacities of action, including the employees' abilities to resist.

However, in spite of the discovery of the growing role of agency in the world of work in knowledge societies, another leading assumption about the labor market that remains unaffected to this day are that skills the individual is expected to bring to the world of work are those which the pre-existing system of work demands and imposes. In an even more general sense, Pierre Bourdieu (1973:72) describes the assumptions (or laws as he calls it) that govern such asymmetric social relations as the tendency of structures "to reproduce themselves by producing agents endowed with the system of predispositions which is capable of engendering practices adapted to the structures and thereby contributing to the reproduction of the structures" of labor. In other words, it is a widely shared assumption that growth of the number of knowledge workers in the modern economy occurs in response to the *requirements the world of work,* especially those its modern technological regime imposes. And as long as one holds on to this assumption the productivity puzzle becomes an even more severe enigma. The share of highly paid and highly skilled workers grows yet a consummate payoff is not in sight.

6. THE GROWTH OF KNOWLEDGE WORKERS

Peter Drucker presents a much more surprising and perhaps plausible hypothesis about the reasons for the growth of the number of knowledge workers in modern society. He suggests that the stimulus for the increasing demand for knowledge-based work has to do less with more difficult and complex job skills, changing technological regimes, the growing complexity and specialization of the economy or enhanced functional steering and coordination needs. The growth of knowledge workers has more to do with the substantial extension in the working life span of individuals and the enhanced knowledge with which individuals come to the labor market in the first place. The observed shift toward knowledge-based jobs should not be seen to result from changing occupational preference and choices of new entrants to the labor market, although changes in preferences of job content and ways of achieving job satisfaction certainly are part of the transformation. In the United States for example full scale labor force entry of adult college trained baby boomers coupled with steady retirements of less-educated persons who reached working age during the 1930s and 1940s resulted in quantum increases in the average educational endowment of the labor force. Prime working age participants with four years or more of college training rose by 64 percent to 25.5 million during the 1980s and those with one to three years rose by 58 percent to 20.8 million (Wetzel, 1995:60).

If we follow Drucker, it is not so much the *demand* for labor and particular skills as the result of more complex and exacting jobs, but the *supply* of highly skilled labor that underlies the transformation of the world of work. More specifically the "direct cause of the upgrading of the jobs is ... the upgrading of the educational level of the entrant into the labor force" (Drucker, 1968:279).[10] Whether the transformation of work in a world of knowledge work by knowledge workers constitutes a historically unique development or will continue to be major attribute of work in the future is an open question. It is likely once the world of work has been thoroughly transformed that "demand" attributes will become more prominent attributes influencing the texture of the labor market. [11]

The much more common perspective is that rapid diffusion of information and communication technologies has altered the production process and the delivery of services and that the employment structure has changed as a result. The new technologies increase the demand for highly skilled workers. Put more technically, most observations "model changes in workforce skill as a function of changes in industry capital intensity and industry-level investment in computer equipment" (Doms, Dunne and Troske, 1997: 254).

At the present time, there are no empirical studies, to my knowledge, that explicitly examine the relationships stipulated by Drucker. However, two recent intensively discussed and researched economic issues that arrive at largely unanticipated results may offer at least an indirect measure of confirmation of the Drucker's thesis. First, a large volume of empirical studies of the US labor market triggered by the observation that income inequalities between well-paid and more poorly paid jobs have risen in recent years considerably, prompted economist to ask what developments may be responsible for these income trends. In particular, the relations between technological change, skill level and income have been studied. The primary assumption examined in these studies is that the increasing polarization of the labor market may be linked technical changes that in turn cause firms to hire more highly skilled labor (Gottschalk, 1997). Second, there are research designs directed toward analyses of the causes of the "productivity puzzle" that could be relevant.

For the researchers concerned with the growing polarization of income levels, two explanations are of particular interest. First of all, the growing differentiation of pay for skilled and lesser skilled labor may be caused by technological change. More precisely, the increase in *demand* for skilled labor and a growing proportion of skilled workers in the labor force is seen as induced by technical developments (Johnson, 1997). The second explanation has a family resemblance to the first account: the *demand* for technologically more sophisticated products and services has triggered a growing need for a highly skilled work force (cf. Bernard and Jensen, 1997:5). In short, changes that can be attributed in one way or the other to demand induced forces provoke a change in the balance of skilled to lesser skilled workers. And, as a result, these transformations in the nature of demand activate and accelerate a growing inequality in incomes by skill levels.

In a broadly-based, cross-sectional, empirical study at the level of individual manufacturing firms and using individual rather than aggregate data for the American economy, Doms, Dunne and Troske (1997) for example have examined the relationship between technology use,[12] the education, occupation and wages of the employees in the manufacturing sector. As the data reproduced in Table 1 show, one is able to conclude that there is a growing covariance between the degree of technology use in firms, that is to say, the progress made in automating both the development and the production process, and the educational level of the employees. The conclusion therefore not only is that "skilled workers and advanced manufacturing technologies are complements" but also that the proportion of employees "in skilled occupations rises significantly with the number of technologies employed" (Doms, Dunne and Troske, 1997: 261, 263; see also Berman, Bound and Geliches, 1994) *as well as* the proportion of employees not directly active in the production process.[13] A variety of controls confirm these findings. In addition, the authors report that employees in firms with extensive technology deployment earn higher wages and salaries.

Table 1 : Technology Use, Education, Wages and Occupation at U.S. Manufacturing Plants, 1988-1990 [a]

	Percent					
	of workers with at least a college degree	of nonproduction workers with at least a college degree	production workers with at least some college	of workers in managerial, scientific, engineering, or precision-craft occupations	of non-production workers	of total wages paid to nonproduction workers
	(1)	(2)	(3)	(4)	(5)	(6)
Plants using less than 4 technologies	9.4	24.1	21.2	33.7	32.7	41.0
Plants using 4 to 6 technologies	12.2	31.2	24.2	35.6	33.3	42.2
Plants using 7 to 8 technologies	14.0	34.5	27.1	36.6	34.9	39.5
Plants using 9 to 10 technologies	16.2	34.9	27.7	37.4	40.7	46.2
Plants using 11 to 13 technologies	15.2	37.5	29.7	33.1	34.3	37.3
Plants using more than 13 technologies	33.1	53.9	34.9	48.6	56.9	62.4
Full sample	18.3	40.1	27.9	38.5	40.5	47.2

[a] There are 3251 workers in the sample that work in plants using less than 4 technologies, 4690 (4-6 technologies), 6403 (7-8), 5914 (9-10), 5931 (11-13) and 7844 workers in plants using more than 13 technologies. Source: Doms, Dunne, and Troske 1997: 262.

The cross-sectional data however cannot offer an answer to questions about the timing of the observed marked substitution in favor of skilled labor in the manufacturing firms that are technology intensive. As a result, Doms, Dunne and Troske attempt to extend their analysis by relating the utilization or adoption of different technologies over time in these firms to changes in the different variables such as the wage levels, the proportion of employees not directly involved in the production process etc. Aside from methodological problems such a procedure may have as the result of the absence of valid longitudinal data, the overall result of their efforts to operationalize "technology adoption" is that "technology adoption is relatively uncorrelated with the changes in the nonproduction labor share, workers wages, or labor productivity" (Doms, Dunne and Troske, 1997: 277).

One possible "explanation" for the "negative" finding would be that the firms the author included in their study already employed or hired a large number of highly skilled employees *prior* to the adoption of new technological means. Thus, "if plants that adopt technologies have more skilled workforces prior to adoption, then we would expect that the pre-adoption wages and labor productivity should be correlated with future technology use" (Doms, Dunne and Troske, 1997:277). The results of their study once more are far from transparent. The author sum up the relation they are able to document as follows: "Plants that adopt a large number of new technologies have more skilled workers both pre- and postadoption" (Doms, Dunne and Troske, 1997:279). As Drucker assumes, one cannot preclude the possibility in other words that highly skilled employees force the modernization of their work places, in the first instance. It is likely therefore that the supply of skilled workers rather than the demand for workers with such skills that constitutes the motor of the rapid and radical transformation of the world of work.[14]

A further but confirmation of Drucker's thesis of a supply induced transformation of the world of work is aggregate data about the growing "skill level" of the population in most OECD countries. According to Johnson (1997:42), the relative skill supply measured as a ratio in the population of high school to college equivalency in the US has risen from .105 in 1940 to .496 in 1993. In five decades, the proportion of the population with a college education has increased fivefold. The increase of college educated labor is particularly strong in 1970s, reflecting a growing proportion of college students in the latter half of the sixties. It would be far too simple to suggest that the tremendous increase in the collective skill level of the work force simply occurred in direct response to market forces. Although individuals will respond to perceived market opportunities, the fit not only in terms of time between perceived market opportunities and education could hardly be expected to be exceptionally close. Too many other factors and forces impinge upon those choices and after many years of education perhaps result in "higher skill levels."

In the modern economy, knowledge is the most important resource. As a result the production of knowledge and learning are the most significant processes in the knowledge society. Public policy in turn must be attuned and attend to these features of modern society (cf. Alexander, 1997).[15] Knowledge societies are changing with rapid speed. For this reason alone it is not sensible to adopt a strict demand or storage model of the kinds of skills and competencies that schools, universities and other educational institutions ought to deliver. Future occupational requirements of the world of work and their obsolescence are difficult if not impossible to anticipate or predict. The close even intimate linkage between prior curriculum and subsequent occupational tasks that is often expected and demanded cannot be accomplished. A decisive feature of the labor market in knowledge societies is its unpredictability and the insecurities that the substance of educationals needs can be determined by future features of the labor market and the world of work. The storage model – schools and universities supply those skills and competencies that can be immediately utilized at work –has to be replaced by a model that couples work and education under conditions of uncertainty and agency.

7. OUTLOOK

Economic theories and research that deal with the productivity paradox for the most part continue to be linked in a rather close fashion to the production process of industrial society, its organization and outcomes. Moreover, the lingua franca of economic theories remains tied to an imagery which displays agents as inflexible creatures entangled in single-purpose, dedicated structures of purposes which imposes its rhythm, often over generations, on passive individuals and groups. As long as economic organizations and regimes such as the labor market are reconstructed as inflexible and largely one-dimensional, the gradients of inequality will be steep and the effects of its logic long lasting. Discourse on the role of economic agents will be concerned with the coercion, that is, the constraints, the vulnerability and effective impotence of individuals and groups in the face of power exercised by the powerful. But new realities require a new language of the world of work. The new language should, in contrast, stress agency, malleability, flexibility, multi-purpose resources, volatility, heterogeneity etc. of social structures and the extent to which individuals and social groups as embedded in their relationships with others have the capacity to employ and transform these structures, once the generalized vulnerability to the forces of the conventional logic of economic conduct has been reduced to a significant degree. The condition for the possibility of greater and more broadly based agency is knowledgeability or a bundle of more widely accessible social competencies and their impact on social structures of inequality accelerating chances for actors to re-fashion social constructions.

Notes

1 The reasons Castells (1996:468) offers in support of his label of network society for the emerging social structures throughout modern society refer to a historical trend in which the "dominant functions and processes in the information age are increasingly organized around networks. Networks constitute the social morphology of our societies, and the diffusion of networking logic substantially modifies the operation and outcomes in processes of production, experience, power, and culture." For a critique of aspects of Castells theory of society see Stehr (1999).

2 I will only cite one more example of the by now rather orthodox claim in science, in education, the economy and in much of the public realm that there is a technology driven demand for highly workers. That claim also resonates strongly with technological determinism: Davenport (1997:2) deferring to Peter Drucker's ideas on the knowledge society explicates the term knowledge-based economy and indicates that the characteristic technological basis of the knowledge-intensive economy, namely information and communication technology, biotechnology, and new materials <u>have created</u> a "remarkable demand for highly educated workers, not only to advance and manage the technologies themselves, but to serve as experts in the finance, production, and marketing of the new products and services which the technologies produce." What is remarkable about Davenport's assertion is that Drucker is one of the few economists who has indicated that the really intriguing dynamics of growth of knowledge workers is the extent to which such gains may not be demand but supply side driven (see Drucker, [1968] 1992:279). I will return to this point in detail later in the paper.

3 Castell's (1996:78) expresses the suspicion that the poor validity of the economic statistics might be responsible for the productivity puzzle and therefore may not even be real: "It may well be that a significant proportion of the mysterious productivity slowdown results from a growing inadequacy of economic statistics to capture movements of the new informational economy, <u>precisely because of the broad scope of its transformations under the impact of information technology and related organizational changes</u>." However, he does not indicate how one might be able to specifically "heal" the deficiencies of the current statistical regime.

4 Jorgenson (1997:4) sees the productivity paradox as arising from the prevailing identification of "productivity growth with technological change". Technological change and productivity gains are distinct. Productivity growth is but a minor component to growth. Technological change occurs, he argues, as a result of investments; economic growth also is due to capital investment. Capital investments can be categorized into investments into tangible assets, human and intellectual capital. The purchase of computers constitutes an investment into tangible assets. But the key concept in this context, intellectual capital remains but a vague and perhaps even more irritating to economists an unmeasured and unmeasurable concept.

---

[5] My cautious observations about a gradual rather than an abrupt, decisive transformation of the economy of industrial society into a knowledge-based economy that fully displaces and renders obsolete the earlier economic formations resonate with Werner Sombart's ([1927] 2000) reflections about erroneous expectations that economic systems can change in dramatic even violent ways: "All those opinions are mistaken which expect a violent upset of the existing economic constitution and a sudden change of the bases of economic life. This opinion too misjudges the nature of economic development, which always proceeds in the form of a gradual, 'organic' reshaping of existing conditions. A new economy 'grows', like a plant, or an animal. Forcible interventions may well destroy, but they build nothing. All previous history confirms the accuracy of this observation."

[6] It is worth noting that the meaning of the term's work and labor now are used almost interchangeably. Labor in contrast to work may or was seen as based in circumstances in another person's direction and control while work refers to what directly maintain one's existence and is carried with relative autonomy.

[7] Several authors have adopted the term "neo-Fordism" to describe the convergence of information technology and the managerial orientation of "Fordism" (cp. Massey, 1984; also Jaeger and Ernste, 1989:).

[8] A brief description of the emergence, nature and recent challenges to mass production in the automobile industry in North America and Europe described as Fordism may be found in Dankbaar (1995).

[9] The so-called scientific management principles that Taylor developed were, for example, designed to limit and restrict the exceptional control workers often had with respect to forms of knowledge that related directly to technical and performance-related knowledge on the shop floor.

[10] Among the surprising even amazing properties of the transformation of the labor market is that the American economy was able to "satisfy the expectations of all these people with long years of schooling.... As a result of the change in supply, we now have to create genuine knowledge jobs, whether the work itself demands it or not. For a true knowledge job is the only way to make highly schooled people productive...That the knowledge worker came first and knowledge work second –that indeed knowledge work is still largely to come -- is a historical accident. From now on, we can expect increasing emphasis on work based on knowledge, and especially skills based on knowledge" (Drucker, 1968:285).

[11] John Kenneth Galbraith dismisses Drucker's argument out of hand. He affirms the orthodox view about the relation between education and the labour market and considers Drucker's perspective as evidence for the typical self-complacency and pretension of the educator misreading the real power balance in society in the process. Galbraith (1967: 238) suggests it is the "vanity of educators that they shape the educational system to their preferred image. They may not be without influence but the decisive force is the economic system. What the educator believes is latitude is usually latitude to respond to economic need." In other words, Galbraith insists that the demand-side explanation generally favored by economists (as well as employers, educators and educational policy makers one should add) primarily accounts for the increase in skilled work.

[12] The degree of technology intensity in individual firms was measured by the authors of the study by counting the number of technical processes or devices such as computer driven machines, robots and so on found in the plants (see Doms, Dunne und Troske, 1997: 287-288 for a detailed description of the different processes and devices).

---

[13] In concrete terms, "the positive relationship between technology use and the percent of skilled workers is primarily due to a dramatic increase in the percent of scientists and engineers in the most technogically advanced plants" (Doms, Dunne and Troske, 1997:263).

[14] In response to the question of the reasons for the immense growth of the service sector in recent decades, Landauer without referring to Drucker, also offers an account that stresses factors induced by the demand for jobs. Thus, new jobs were needed, so new services were invented. Many new or expanded services depended on computers: a plethora of investment instruments -- complex new mutual funds and trading schemes, a deluge of new insurance policies and options, a myriad of debit and credit cards, dozens of new kinds of bank accounts and novel banking services offered from widely dispersed branches and machines, multitudes of new medical techniques and therapies, fast food restaurants, fast copy stores, fully filled planes with frequent flyer plans, mom and pop mail order firms, direct marketing, PC maintenance, and so forth (Landauer, 1995:74-75).

[15] Concerns that the quality of the available jobs may not be compatible with rising educationl levels (Harman, 1978:209) correspond to exactly the opposite perspective, namely that the quality of the world of work is primarily driven by the nature of the demand.

Chapter 13

Empowering Information and Networks through Adaptive Public Policies

JOHN DE LA MOTHE[1]
PRIME, University of Ottawa

"The art of association becomes the mother of action, studied and applied by all." Alexis de Tocqueville, *Democracy in America* (1830)

"A society composed of an infinite number of unorganized individuals...constitutes a veritable sociological monstrosity." Emile Durkheim, *The Division of Labour in Society* (1933).

1. INTRODUCTION

It is commonplace to observe that 'the information revolution' is having profound effects on all aspects of advanced industrial life. It is widely accepted (albeit not uncritically)[2] that after 'the end of history' (Fukuyama, 1992) we have now reached an era in which tangible products are being supplanted by information goods and services – from music and movies to software and stock quotes. (Shapiro and Vrian, 1998) Institutions, individuals and governments are being forced to deal with the changing nature of work, learning, and trade, as well as with their own changing expectations. Competition and careers, culture and public policies are all apparently up for a major re-think. All are being re-structured by information and information and communications technologies (ICTs). It is therefore not surprising that terms like the Information Society and the Knowledge-Based Economy have entered so easily into popular currency.

2. THE NEW LANDSCAPE

However the debates about restructuring have in fact *not* led to a widespread re-think. They have instead relied on antiquated frameworks and have been largely relegated to reinforcing issues related either to the ways in which goods and services are developed, designed, produced and distributed or to the globalization of technology, competition and finances. (Ruigrok and Van Tulder, 1995) The clear winner here has been not free trade (with all of its best Keynesian, post-Versailles intentions) but the free market. This, it is claimed by the geo-strategist, Edward Luttwak, is a bit of short-hand by which is meant more than the freedom to buy and sell. "What [free marketeers] celebrate, preach and demand is private enterprise liberated from government regulation, unchecked by effective trade unions, unfettered by sentimental concerns over the fate of employees or communities, unrestrained by customs barriers or investment restrictions, and molested as little as possible by taxation." (Luttwak, 1999) Seen from this one extreme there is both a deliberate ideological move to encourage a retreat of the state from above (effectively vacating the *polis*) *and* a widespread and active facilitation by ICTs to erode the state from below. Thus we have a hollowing out scenario. But as Samuel Huntington and others have articulated, (Huntington, 1996; Rosenau, 1997) while information and ICTs are profoundly re-shaping democracy, public management and public policy, they are at the same time driving up and re-defining the public demand for government. And yet there is only a tangential focus in public discourse on the need for adaptive public policies. Hence, frameworks for analysis have remained macro, traditional and reactionary. (Doern, Pal and Tomlin, 1996) We are left drawn either into the techno-euphoria camp in which social ills and wishes will be solved by technologies meeting market demands (realized either in terms of price mechanisms or articulated in terms of technology fetish and fashion) or we are drawn into techno-pessimism in which de-skilling, displacement, and exclusion are the norm.[3]

Regardless, what the information revolution *has* spawned is a broad-based lexicon of transformation. Terms like 'virtual community', 'cyber-space', and the 'infobahn' have been deployed by media consultants, e-marketeers and political spin doctors so effectively that the public has been reinforced in its own unnerving sense of alienation, 'millennialism' or, simply, of 'change'.[4]

This is evident at both ends of the spectrum. In North America, for example, there is social concern over the growing economic gap between the top 10% of society and the bottom 10%,[5] of the so-called 'disappearing middle class' and of the information 'haves and have nots'.[6] There is concern over the uncertain professional futures of Generation Xers as well as the changing nature of telework and careers. (Halstead 1999) At the same time there is widespread enthusiasm for the potential that waits to be harnessed through telelearning, telemedicine, teleworking, and so on. There is a sense of expectation associated with the fact that more than 70% of new job creation in many advanced countries is in the information-based services sectors. Together these issues and perceptions are coupled with the strains popularly associated with free trade and globalization alluded to above (such as value added competitive advantage, high technology exports, technological balance of payments, the dematerialization of trade, national and regional interdependencies, and so on). (de la Mothe and Paquet 1996) All are associated centrally with the pervasive role of information and ICTs.

Evidence is suggestive. Government expenditures on ICTs as a percentage of GDP rose to nearly 8% in both the US and Japan while the average in the EU and the OECD respectively reached nearly 7% between 1992 and 1997. Governments anxiously measure the PC penetration in households, PCs per 100 white collar workers, the number of internet hosts per 100 inhabitants, the cost of internet access, innovation in the ICT industries, and so on.

The combined effect is putting public policy seriously to the test as it struggles to cater *both* to issues of social inclusion and cohesion while at the same time promoting the growth of high technology multinationals and inward-bound technology intensive foreign direct investment.

The crisis of governability, which arose in the 1960s, and the shift towards conceptions of new governance are all part of the emerging 'cyber-society'. We are all walking on a new landscape and we need new frameworks of action-analysis.

But what is important about this 'information revolution' is not simply that information is ascendant as a determining factor in sustained economic growth and activity. This has long been true. (Hayek 1945) Instead, it is the sheer pervasiveness of ICTs; the convergence of ICT capabilities worldwide; the speed of interaction, transaction, and transmission; and the storage capacity for various types of information that has driven the aggressive re-shaping of institutions and economies, norms and practices alike, a point developed by Castells (1996). However, the lenses through which we view these transformations are opaque and need to be cleared and re-focused.

This chapter takes a broad view of the ambient shifts that are influencing the direction of policy choices and issues. It will briefly review some prevalent macro-socio-economic reactions to both Technology and Government which serve to reveal reactive assumptions about these terms and their effects as well as to largely obfuscate the dynamics of the contemporary picture. It will then re-focus the present context in terms of complexity, flexibility and networks, and will thus adjust the unit of analysis away from traditional interventionist views of public policy, outcomes and structures and towards processes and ligatures. As a result it will sketch some issues and choices that policy makers need to attend to, and will close with a few modest proposals on how to empower information networks through adaptive public policies.

3. MACRO SOCIO-TECHNICAL FRAMEWORKS

In thinking about information technology in general, especially in the context of the re-shaping of public policy, it is not surprising that many broad frameworks are principally predicated upon one's own fundamental emotional reaction both to 'Technology' and to 'Government'. Neither term is exempt from prejudice. For example, Robert E. Lane, who was one of the first authors to employ the term "knowledgeable society" in 1966, tied his framework rather closely to the promise of a particular conception of science and technology that reflected the optimism of the early 1960s. This view found its counter-point in the critical theories of Herbert Marcuse (1964) which skeptically saw technologically-mediated society as producing a 'one dimensional man'.

In *The Age of Discontinuity* (1969), Peter Drucker not only highlighted the centrality of knowledge in socio-economic relations but he also captured a certain angst with his title. This angst was prevalent in many post-1968 social movements in North America, France, the U.K. and Germany which were deeply anti-technological (luddite), anti-MNE (counter-culture) or anti-market (neo-Marxian) in bias.[7]

Daniel Bell, in his *The Coming Post-Industrial Society* (1973), identified knowledge and information as being fundamental resources. Indeed, he noted that the post-industrial society is a knowledge society for two basic reasons: (1) "the sources of innovation are increasingly derivative from research and development (R&D) – and more directly that there is a new relation between science and technology because of the centrality of *theoretical* knowledge"; and (2) "the weight of the society – measured by a larger proportion of Gross National Product and the larger share of employment – is increasingly in the knowledge field". (Bell 1973, 174) These terms would clearly be recognized today.

Thus, this broad set of realizations and debates between technological utopians and dystopians regarding the growing importance of information and the ways in which information and technology are produced, stored, owned, diffused, and consumed, were widely discussed and advanced through such neo-liberal assessments as John Kenneth Galbraith's *New Industrial State* (1967) in which he deposited a new employment class – the technocrats – into the popular lexicon (what Robert Reich would today call 'symbolic analysts') (Waligorski) as well as through more socialist perspectives as those offered by Radovan Richta (1977) and Helmut Schelsky (1992).

Throughout, macro-social images of technology and its effects have been strongly implicit, both in the analytic literature and on the popular movement fronts. Sometimes they have been explicit. Francis Fukuyama's recent book on "The Great Disruption" (1999), for example, posits that a dynamic technologically innovative economy will by its very nature disrupt existing social relations. Along with the widespread Blakean Luddite (in Fukuyama's case, neo-Darwinian), anti-corporatist counter-culture and resilient neo-Marxian *social* perspectives -- which were predicated on assumptions of technology-led social exclusion, 'massification' and codification in the style suggested by Walter Benjamin in his classic article on "Art in the Age of Mechanical Reproduction" (1923) -- there was also a popular categorization of knowledge production in terms of disembodied (a-social, a-political) 'technological imperatives' in which technology would develop inevitably according to its own inner logic. This view, which was crystallized in the popular mind through such terms as 'fool proof' systems developed for security contexts (in which the 'fools' were humans) and such film characters as Stanley Kubrick's 'HAL' in *2001* (a computer which over rode human decisions and which sardonically was only one letter short of 'IBM', thus making both a technological and a corporate statement), saw the advance of technology and technological systems as being cybernetically natural, relying on no human impetus, input or guidance. But, as Nico Stehr and others in this volume (and elsewhere) (Biker and Law 1992) have clearly demonstrated, technologies are socially created to meet social needs. They are socially constructed. Indeed, as the great physicist Erwin Schrödinger put it in 1952, "there is a tendency to forget that all knowledge [science, information] is bound up with human culture in general, and that findings, even those which at the moment appear the most advanced and esoteric and difficult to grasp, are meaningless outside their cultural context."[8] This is a point to which we will return.

Nonetheless, the collective responses to information and ICTs are often predicated on our *a priori* views of 'Technology' -- such as 'progress is movement for its own sake'[9] -- and thus do not lead easily to clear thinking on how information and ICTs are to be promoted and harnessed, socially or economically. Similar problems cloud our reflections on the changing nature of government and public policy, particularly in the realm of information, innovation and impacts.

4. CRISES IN GOVERNABILITY

As *The Economist* detailed in 1999,[10] even though the world has seen a rise in the number of new democracies since the collapse of communism, citizens in such mature democracies as Canada, the US, Germany, Australia, Finland and Italy are losing confidence in their political institutions. Indeed, measured along three axes – the honesty of leaders, the perceived caring of politicians and trust in government – citizens in 11 advanced countries have registered a steady decline on all three fronts since the late 1960s. Moreover, in many countries, voter turnout for national or presidential elections is at an all time low. Thus, to many, government is under siege. Governability is in crisis. This perception is manifest on at least three levels.

Politically, and using traditional frameworks, governments seem to have run out of ideas and their authority is being challenged, more by the pace of events and by the pervasiveness of interdependence than by any one source. In response, the traditional dichotomies between left and right have been made moot by the rise of amalgam politics, sometimes associated with 'the Third Way' in Britain and the United States. (Giddens 1994, 1999; Keane 1998) The collapse of the post-Keynesian consensus which allowed temporary deficit incurment and which accepted 'big government' programs has yet to be convincingly replaced with a new vision that is appropriate to the time. Some analysts have rightly argued in their defense of government that the public trust must be regained. (Weisberg 1996) Such social capital observations are indeed valid but, left on their own, they are operationally vapid. Other analysts speak of the need to 're-invent government'. This too is instinctively correct given the new context, but again it lacks precision.

Administratively, the downsizing of government spending throughout the OECD member countries has led public service managers to not only try to do more with less but also, and more importantly, to try to find new ways of doing things. The language is telling as public administrators now speak of 'breaking through bureaucracy', providing 'value for money', citizens have become 'clients' to whom 'services' are to be provided, and so on. However, not only has the individual been commodified in this process and thus reduced in value as a citizen who is to be represented by their elected officials, but in a context in which sovereignty in being perforated by a laissez-faire, market oriented, globalized economy, facilitated and driven by ICTs, the self-determination of people is widely perceived as being lost. (Kuttner 1996) As we watch the declining authority of the state, as Susan Strange has put it, we watch too the shift away from politics to new patterns of power in which the neglected factor is technology. (Strange 1996) Moreover, we have adopted the language of commerce to re-define government, and this is insufficient.

Organizationally, in a context of interconnectedness, vast flows of information, the de-materialization of trade, investment, and value-added activities, and so on, the hierarchical organization in which decisions are made from the top down – as in government - are ineffective in the extreme. Taking a page out of corporate restructuring, large multinationals have learned to devolve their decision-making processes down to the lowest appropriate level. They have re-organized themselves into 'federations of firms', breaking the large ocean liner with many business units and which is hard to turn or stop into a fleet of speed boats which is responsive, communicative and innovative. (de la Mothe and Paquet 1994) Of course, perhaps nothing is as information intensive as government, but the lessons of corporate restructuring are not easily applicable to public administration. This is so because first, corporations tend to have a unified goal and tend to know who their clients are. Government has myriad goals and multiple clients. (de la Mothe 1999) Second, government departments tend to be more territorial and political (reporting as they do to Ministers or their equivalent) when it comes to horizontal coordination or the possible re-distribution of their resources.

Nonetheless, together these dilemmas reveal a serious need for *rapprochement* if we are to recover an effective analytic lens for public policy in the context of information and ICTs. The essential problem was quietly foretold in 1970 by Daniel Bell when he said that 'government [today] is too big for small problems and too small for big problems'. Many citizens feel this.

One response to this erosion of faith in government has been to embrace the truism issued by former Speaker of the House Tip O'Neal who said that 'all politics is local'. This seemed like a rationale (as if one were needed) for citizens to rally around local causes and issues. California and Sweden are classic examples of the use of public referenda to settle a plethora of public policy problems, from school closings and environmental regulation to industrial re-zoning and animal rights. It may have started with citizen protection groups such as 'Nader's Raiders' around automobile safety and the environment in the 1960s or with the powerful role that non-governmental organizations – from Amnesty International and Greenpeace to international organizations and citizens groups – have come to play. At one extreme has been the seductive emergence of a powerful commitment, particularly in the United States, to communitarianism which gnaws at the *raison d'être* of central government while at a more diffused end of the spectrum has been a sense that the rising power of citizen groups resulting in what some are heroically (albeit erroneously) calling 'a new liberalism'. (Berry 1999)

These various reactions all serve to tell us that recognition of change is widespread. But the frameworks which are chosen tend to be those which are ill equipped to assess the new landscape. It is not just that they are too macro. Indeed, as noted already, Susan Strange has highlighted that both technology *and* capitalist democracies are neglected elements. If one were astute and historically oriented, then one would likely accept Francis Fukuyama's initial assessment of social dislocation. Building on his core thesis in *The End of History* (1992) and his conclusions in *Trust* (1995), he locates his reasons for widespread social dislocation -- the arrival of the information era, the changing nature of employment, the culture of individualism, the birth control pill -- in terms of technology and the loss of social capital. (Here he is not original since de Tocqueville did much the same in 1830 and 1835.) Hence the erosion of informal rules, norms, and trust. For Fukuyama, without social capital it is clear that there can be no democracy, and with that no global capitalism. Thus in his terms it would seem natural for techno-economic imperatives to forcibly forge their own social norms (in the sense suggested above in terms of technological imperatives). Of course, the prophet of this pathway is Joseph Schumpeter whose notions of 'creative destruction' in the capitalist universe clearly extended to the realm of social relationships. But where does this lead us?

5. FROM GOVERNMENT TO GOVERNANCE

In broad terms so far we have suggested that information and ICTs are having a profound effect on the shape of democracy, government and public policy. We have also, without delving into any of the specific dynamics involved in developing or deploying new information-based products and technologies, suggested that macro-socio-economic frameworks which have evolved since the 1960s have reflected the *zeitgeist* of our time *viz. à viz.* Technology and Government, but have failed to offer any prescriptive analysis of use to policy makers. The choice left by such a vacuum can lead to the abdication of responsibility for the deployment of ICT and lead directly into arguments about 'the end of government', the loss of governments' legitimation and authority, (Habermas 1974) or the 'merits' of bureaucracy taking on market models as a substitute for the public or national interest. (Clinton 1994) A large part of the problem rests in the unit of analysis used for our frameworks. By imagining our context as being a battlefield of competing interests and institutions, we quickly erected scenarios in which flexibility and responsiveness were foreign. In that sense we have not overcome de Tocqueville's classic problem of modern democratic systems: i.e. the necessity of securing citizens' participation and responding to societal demands while being efficient in their problem solving.

This is a core issue in an information society, and the response has been a rise in 'governance' as an approach. Over the past fifteen years, governance is one of the vogue terms that has emerged. Its difficulty lies in the fact that it is not 'government' but it is not clear exactly what it means. R.A.W. Rhodes, for example, has delineated six distinctly differing uses of the term, from being a code word for the minimal state to corporate governance. (Rhodes 1996) In the former sense, "governance is the acceptable face of spending cuts". (Rhodes 1996, 653) In the latter sense, "the governance role is not concerned with running the business of the company, *per se*, but with giving overall direction to the enterprise". (Rhodes 1996, 654) Neither is especially helpful.

More usefully, as Ali Kazancigil (1998) has traced, the notion of governance initially emerged from the urban context in the beginnings of the governability crisis (noted above) to cope more efficiently with local problem solving. More recently it has been applied to central government operations. These contexts and responses were related to the growing complexity of issues that policy needed to face and to the growing plurality of agents in civil society, particularly those involved in the policy process. It was assumed that by increasing the society-state interaction at a time of government down-sizing, communications and efficiencies could go up thus ensuring a solid continuity of effective policy making. (Gore 1994)

While loose in detail, the governance approach to analysis opens the door to the greater capacity to cope with policy making issues in increasingly differentiated advanced modern societies where the various networks and sub-systems have become more autonomous. (Kooiman 1993) Such a development clearly puts governmental institutions and instruments under strain. But, embracing a framework which highlights the interaction of players, the co-evolution of outcomes, the sharing or negotiation of decisions, the lowering of power to the lowest appropriate (i.e. effective) level, the plurality of interests, and the role of information and knowledge, all improve the analytic and operational potential of policy in an information environment. Moreover, it allows us to re-frame public policy away from intervention in which governments 'pick winners' (in the sense of traditional industrial policy) towards one in which government 'backs leaders', that is – is a partner and a facilitator. The leader becomes animateur. (Paquet 1999)

There are of course two dangers that need to be protected against in a governance framework. (1) The weakening of the hierarchical dimension of public management, and the emphasis on market and coordinative modes of managing resources, reduces the possibilities of the central government to implement long term policies on major social issues. (2) And by working in a negotiated context, it is possible to view policy making as being politics-free and thus runs the risk of obscuring the essential function of decision making and legitimate responsibility. (Paquet 1999)

The capacity to govern or formulate effective public policies in an information-intensive context will not of course improve automatically by moving from an 'old government' to a 'new governance' framework, since it will rely on how well it incorporates representative and participative politics. But if it succeeds in so doing, then it would be well suited to handling the complexity of the information society.

6. ADAPTIVE PUBLIC POLICIES

In dealing with both complexity and information, I have elsewhere argued with Gilles Paquet (1994; 1996; 1998a; 1998b) that the issue lies not in focusing on outcomes and structures but on processes. This has been reflected above in the critical review of macro-socio-technical frameworks that led to an inappreciation of the crisis of governability. Indeed it is becoming widely recognized by policy makers that in an information society typified by the dynamics of techno-globalism that institutions and policy instruments must increasingly be made flexible and adaptive. Adaptation is a dynamic mechanism, based on learning, that depicts the adjustment by actors or groups (and absorbed – tacitly or explicitly - by their operational protocols and sub-routines) to their environmental (social, technical, political or economic) constraints. Through many ongoing rounds of information exchange and interaction, the overall system as well as the individual actors evolve, re-position and are transformed. The actual pattern of transformation depends partly on the capacity of the individual, institution or instrument to be engaged meaningfully in its 'cognitive space' or what Barry Wellman and his colleagues have referred to – in McLuhanesque fashion - as 'the media space'. (Mantel, Baecker, Sellen, Buxton, Milligan and Wellman, 1994) Here we can sense the full breadth of knowledge, from information and data to creativity, insight and judgement or wisdom, which together forge social capital.

This allows us to view innovation, in the sense spelled out by Schumpeter's 'creative destruction' that energizes capitalist exchanges, as new practical knowledge. It allows us to see individuals and institutions (reflected in their instruments) in terms of their capabilities to explore and exploit new possibilities (and thus to make strategic choices). And finally it allows us to examine the features of the broader social processes within which organizations are nested. Taken together, we can begin to discuss the governance of communities of practice in the sense that Peter Haas refers to 'epistemic communities'. (Haas 1992) While formal 'glue' for such cognitive spaces can be found in such stability and continuity enhancing rules as law, in the dynamic information society the more meaningful glue is to be found in cultures, communities and networks. This is important for policy consideration given that information and ICT are widely held to erode or destroy culture and community.

Of course, sociologists who have defined community and culture in terms of *place* have long fretted that erosion was inevitable due to bureaucratization, industrialization, urbanization, capitalism, technical change and so on. But by re-orienting the assessment to focus on social networks and relationships, then in a world of the 'global village' – as McLuhan originally said – one's 'village' could span the globe. Thus community is re-defined in terms of *space*. (Wellman and Guila 1998; Garton, Haythornthwaite and Wellman 1999) Through such a lens, computer networks that connect people and organizations are quite capable of creating and transforming in a cohesive way the nature of community and culture. Thus 'community' can be expanded to include expertise networks as well as computer mediated communication (CMC), computer-supported social networks (CSSN), and so on. The important cultural element in electronically mediated social networks is the presence of inter-personal cues. For example, a video game might provide the user with an a-social stimulus. However, in a computer network, users establish roles, routines and protocols interactively and thus begin to build cultural norms of engagement and behavior.

These brief emphases on the importance of processes, networks, and adaptability in a complex and information-mediated context are portentous for public policy. As Gilles Paquet (1999) has argued, policy makers face a "moral vacuum". The 'glue' of social capital is critical to energize an information society. Using a governance framework, we can move away from a guiding philosophy of intervention towards a philosophy of public purpose. This would allow for a safe devolution of responsibilities out from the center while ensuring the cohesion of the *polis* (uncovering new partnerships, sharing and blending interests, empowering and expanding citizenship, etc.) through information and ICT.

7. ISSUES AND CHOICES

Since the presidential debate in 1960 in which the age of televisual democracy was born,[11] the transformative impacts of information and ICTs have thoroughly penetrated the realm of public policy. As we have already argued, the role of the public sector is being broadly re-shaped. The power base of decision making is shifting. In a sense, we are collectively caught in an institutional migration away from a policy based on power, institutions and exclusivity (e.g. sovereignty) towards one based on processes, flows and interactions. 'Policy' has been defined in outcome-addicted frameworks as being simply "watching where the dollars go". (Branscomb 1994) More organizationally-jealous approaches have protected the siloed-policy framework of departments in terms of discrete responsibility over policies for investment, competition, telecommunicaitons, foreign affairs, standards, technology, science, and so on. (Doern 1996; Pal 1999) Numerous OECD governments, in an effort to better manage public resources and lower deficits and debt, have attempted to orchestrate horizontal coordination across portfolios. These have largely been top-down efforts, and thus have been flawed in design. We need a different type of public policy if we are to govern the new information-based context.

Thus to overcome such paradigmatic yet still hegemonic conceptions of policy, one might focus on the driver of change - knowledge. To do so one might turn to a definition of 'science and technology policy' that holds that it is "a practical activity concerned with developing, managing and supporting the capabilities of countries to advance and apply knowledge". (de la Mothe and Dufour 1995) Through its focus on knowledge, this is a definition that could well be applied to 'information policy', an area that clearly could cross departmental borders, histories and (hopefully) power structures, and thus open the doors to adaptive public policy. (de la Mothe and Paquet 1996)

7.1 Electronic Republics

One policy outcome of the information revolution is to allow for the possibility of an electronic republic. (Grossman 1995) C.B. Macpherson discussed this rather optimistically in *The Life and Times of Liberal Democracy*. But what such a concept suggests is that citizens are feeling powerless, cynical and frustrated about government (our forementioned 'governability crisis'), and that the traditional barriers of time and space have fallen to technology thus leaving our political institutions exposed. At the same time ICTs are also shrinking the distance between the elected and the electors. Thus the opportunity does exist for a more open, participatory democracy in which individuals not only are asked for their input into policy decision making (beyond the banal cooption strategy of some governments for 'public consultation') but they are also able to use the new technologies to create for themselves new communities of interest and practice. (Schlove 1996) Evidence suggests however that with the information explosion via the Internet and other cyber-sources, people (a) do not necessarily want further access and involvement in public decision making, and (b) people are genuinely faced with a problem of determining which sources offer real or quality information. Senior analysts at the Congressional Research Service and the RAND Corp. in Washington DC have said, for example, that if they want to know what's going on they avoid people who spend an appreciable amount of time on the Net and deal directly with people in their offices.[12]

7.2 Cyber-Citizens and Smart-Communities

It is not business as usual. In the Information Society, 'citizenship' becomes re-engaged with responsibilities – moving up *and* down the decision system and carrying with it new tacit knowledge of system dynamics, organizational tendencies, etc.. Through existing 'connectivity agendas', many governments are actively engaged in a race to 'plug' their communities in. The Canadian Government, for example, defines 'connectedness' as "a measure of the overall ability of Canadians to utilize information and communications technologies to interact and transact with one another and the world." (Industry Canada 1999) In undertaking such a task, policy makers have adopted a neo-Keynesian framework, replacing physical infrastructure with 'smart infrastructure' in the hopes that net benefits will accrue. Thus, as chapters by Sciadas, Steeves and Winer in this volume have already revealed, concern over rapidly improving the penetration rates of telephones, cable, home computers, modems, cellular phones and internet across the nation have become the proxy measure of policy success. They are not. Unless a policy initiative deals directly with cultural content and infrastructure compatibility (i.e. it is no use having a 286 in a Java world), then sovereignty will genuinely be eroded in an Ohmaean sense and the socio-economic potential will not be realized. In an era of constructed techno-economic advantage in which Ricardian natural endowments do not matter, this could be a serious mis-step for policy makers. (Freeman 1996) Moreover, offering access to information while charging sufficient fees to discourage requests can suggest an opposite effect.

7.3 Social Inclusion

The issue for policy of the social union is a broad and complex one. There is nothing automatic about the ways in which various groups of differing characteristics will access or respond to new ICTs. In Canada, for example, the province with the lowest use of computer communications is Quebec which raises the question of whether this is a result of language, culture, income levels or a combination of these and other factors. (Dickinson and Ellison 1999) But as technological change races ahead in the liberalized trade regime, there is a practical cost to not closing the information gap. Seniors, the disabled, the disenfranchised, the distanced, all should be targeted by policy for inclusion, not to stoke the global capitalist agenda but more to mobilize the sense of community and thus of engagement.

7.4 Privacy and Information Access

Policy must also come to grips with the absolute necessity for individual privacy. In an age of (imminent) enhanced citizen re-engagement, it is hard to rationalize or sanctify the loss of privacy to anonymous corporate interests (for example, the giving up of credit card purchase and travel patterns, telephone use patterns or of medical information and the subsequent sale of said information to third party retail or surveillance interests). As the recent debates over celebrity pursuit by photographers, policy makers must decide whether privacy is 'given up' or 'taken away'. The popular answer both on this and on the power of the media is fairly clear. But the more complicated answer reveals such critical issues as: (i) How do we structure trust through cryptography? (ii) How can we control surveillance? (iii) How can we protect personal data? (iv) How has privacy been transformed from being a right to being a commodity? And so on. (Agre and Rotenberg 1997) Some prefer market solutions and self-regulation of practice to such matters. Some believe that the new service industries will generate their own 'infomediaries' that become brokers for information between consumers and companies. The latest technologies have also spawned an intense debate between governments over encryption, but even encryption will only re-store internet users to a level of privacy that they commonly expected five years ago. And such radicals as David Brin (1999) have suggested that the solution is to abandon privacy as an issue altogether. Most, however, would shrink from such an idea, recognizing that privacy and secrecy are indeed akin, thus opacity – not transparency – are the desired principles of behaviour.

7.5 Smart Property

A critical element for consideration by policy makers concerns the ownership of information and intellectual property. Increasingly intellectual property will play a critical role in economic prosperity, cultural vitality, national identity and human rights in all countries. Since the 1960s, intellectual property rights have moved from being the bastion of the chemical and manufacturing industries based on demonstrated novelty to being key competitive tools in such fusion fields as biopharmaceuticals, materials as well as informatics. Here, broad-based approaches based on prospect are supplanting novelty as the core issue and product cycle times have driven some leading edge ICT firms into a reliance on industrial secrets rather than disclosure. (Wallerstein, Mogee and Schoen 1993; Doern 1998) However, controversy is not simply a feature of questions over ownership or legal definitions of property. Indeed, the new ICTs and new emerging technologies, coupled with the new world order (geo-strategic), international political economy (geo-economic), trade product mix, and MNE strategies are stretching to the extreme existing IPR regimes which were initially established in the 15th century. Government policy must directly address this, in its national interest *as well as* in the interest of equitable world development. (Cimoli and de la Mothe, 2000)

7.6 Flexible Policy-Flexible Organizations

Clearly one of the great effects of information and the ICTs is the changing nature of organizations, work and expectations. Tony Killick (1995) has taken on the analytic task of explaining the nature of flexibility and adaptability head on. The European Commission has initiated a practical policy discussion on the need for flexibility. (1997) This includes the need to address both the technological and the societal environments, together, in a way we outlined at the outset; the need to improve the innovative processes within organizations; coping with outsourcing, dealing with the contradictory demands of flexibility and security, negotiating change, and the need to facilitate the development of new working arrangements and relationships. (Castells 1996) Paradoxically to the European Commission, but not surprising here in this volume, is the observation that social networks are not being pushed out by technology but are absolutely requisite for an effective information society to be realized.

8. CONCLUSION

The dispersive power of information and communication technologies have long held the promise of opening democracies to public participation and shared decision making. By pushing towards a desegregation of decision making downwards, the possibility has emerged to reclaim politics and public policy at a lower level than the centralized structures and mind sets would otherwise allow. (Kennedy 1999) In other words, political systems, public management and public policy have all been re-shaped by the pervasiveness of ICTs. However, this dream of dispersion has not yet been realized. Indeed, the reluctance of political and managerial structures to share power, coupled with the inherent tendency of technology to channel and agglomerate benefits very narrowly in a socio-economic context, suggest that the adjustment process is not automatic. Indeed, some would say that these tendencies strike at the very heart of the Weberian bureaucratic state and fuel Clause Offe's thesis (1982) on disorganized capitalism. Yet technology's impact on politics, public management and public policy are profound. Adaptive policies and programs that can focus on ICTs need to

- develop social capital;
- distribute information and decision making responsibilities to the community;
- reach into hitherto disenfranchised groups;
- protect individual information;
- define information property in a way that facilitates competition while ensuring reasonable protection of entry barriers;
- improve inter-agency coordination of information, policies, programs and actions; and
- build flexibility both into policies and public organizations.

If government is to regain a core *raison d'être* for the public interest – and thence to provide a real defense of government at a time when there is still a crisis in governability - then it must wean itself away from pseudo-industrial policies which fly in the face of negotiated rules of the WTO and other international bodies. It must view and recapture culture and community as being central (but only understood in the cyber-society sense), and it must recognize the enhanced requirement, not only to produce more technology but, to better manage and understand it. Wired governance cannot attend to only physical infrastructure and macro rules of the game. Social capital, culture, content and community must be addressed if adaptive public policies are truly to empower information and networks.

Notes

---

[1] The comments of Fred Gault and Barry Wellman were most helpful, as was the research assistance of Tyler Chamberlin.

[2] See *The Economist*, "How Real is the New Economy?", July 24-30 1999, p.21-24, p.67

[3] See for example Brian Easlea, *Liberation and the Aims of Science*, Brighton, Belhaven, 1984; Jerome R. Ravetz, *Scientific Knowledge and its Social Problems*, Oxford, Oxford University Press, 1979; Chris Freeman and Luc Soete, *Work For All or Mass Unemployment?: Computerised Technical Change into the 21st Century*, London, Pinter, 1994

[4] This is discussed in John de la Mothe, "Science and Literature in the Age of the Cyber", *Canadian Journal of Political and Social Theory*, 1989

[5] *The Globe and Mail* has reported that the multiple between these social classes has grown from 21 times to 310 times between 1971 and 1996.

[6] Bell companies in the United States have been reported as no longer repairing phone box service in inner cities such as Detroit due to the danger posed to repair crews as well as to the high repair rate. Some neighborhoods have been reduced to 911 service only for the block thus cutting families and friends off from regular communication and changing the shape of community.

[7] In North America, we can point to the cornerstone influence of Rachel Carson's *Silent Spring*. In Britain we can think of the Greenham Common Women. In Germany we can note the rise of Green Politics.

[8] Erwin Schrödinger, "Are There Quantum Jumps?", *The British Journal for the Philosophy of Science*, *II*, 1952, p.109

[9] See Chapter 1 of John Ray, *Enlightenment's Wake*, London, Routledge, 1995

[10] *The Economist*, July 17th – 23rd, 1999, "Politics in Brief, Is There a Crisis?', p.49-50

[11] This first televised debate was between John F. Kennedy and Richard M. Nixon. Attesting to the power of communications technologies, it is noteworthy that those who watched the debate on TV thought that Kennedy had clearly won while those who listened on the radio thought that Nixon had clearly won.

[12] Personal communication with Caroline Wagner, RAND, and Christopher T. Hill, formerly of the CRS now Vice-Provost (Research) George Mason University.

PART V

NEW DIRECTIONS

Chapter 14

Conclusions

JOHN DE LA MOTHE AND GILLES PAQUET
PRIME, University of Ottawa

1. INTRODUCTION

It would seem clear that the chapters in this volume have offered almost as many challenges to statistical agents, policy makers and researchers as information and ICTs themselves do in our everyday worlds. However, there are some commonalities.

Many of the chapters pointed towards the importance of dealing with the dynamics associated with processes, linkages, networks, cultures, communities, knowledge flows and trust. In a sense then, the emphasis has been on meso-level analysis. In an era typified by complexity, speed and the need for flexible systems, it is not surprising that this was juxtaposed with what has to date been typically a preoccupation with inputs and outcomes, with micro and macro.

Some authors have called for new frameworks that would themselves require new approaches to measurement and policy analysis. Again, here, the critical importance of focusing on the comparative dynamics of mediation – on values, attitudes, 'social glue', socio-technical filters, and so on – was echoed as they are central to understanding the success and failure of the adjustment mechanisms and capabilities of institutions, firms, individuals, and government agencies. In this regard, for example, Keith Newton dealt with such items in terms of learning firms.

Numerous chapters served to remind just exactly what it is we are dealing with. In a knowledge-based economy, we were reminded that information is not knowledge. It is organized data. Thus authors consistently cautioned (some quietly and others less so) against dealing solely with the commodification of information. Moving from a preoccupation with information, in and of itself, towards a more utilitarian view of information with an intent of creating, adopting, applying and distributing information for socially useful purposes (e.g. social advancement, social cohesion, etc.) and creating knowledge were strong undercurrents throughout. Indeed, the central importance of social capital and the changing natures of community, culture, and access were seen to be key.

Authors also emphasized that there is no technological imperative. Technologies are not self-generating but are instead socially shaped to meet social needs. Thus government programs, incentives and policies are distinctive in function, as is the inter-locking co-evolution of social norms and practices. Nico Stehr highlighted in his considerations of labour markets and productivity, for example, how skills and adoption of new technologies are variable across institutions and sectors. Valerie Steeves, too, illustrated the role of socially constructed legal frameworks in shaping technologies, including ICTs.

Although students of such varied fields as the social studies of technology or systems theory might not be surprised by these highlights, the nature of the beast poses critical operational and conceptual challenges for those interested in measuring, researching or deploying in an even-handed way the innovation patterns and social impacts of information and ICTs. After all, what is important about this 'information revolution' is not simply that information is ascendant as a determining factor in sustained economic growth and activity. Instead, it is the sheer pervasiveness of ICTs; the convergence of ICT capabilities worldwide; the speed of interaction, transaction, and transmission; and the storage capacity for various types of information that has driven the aggressive re-shaping of institutions and economies, social norms, practices and routines.

2. ORGANIZATIONAL OBSERVATIONS

Each of the authors in this section of the book highlighted the role of interactivity in understanding how information and ICTs permeate and transform modern socio-economies. Interactive learning, for example, was focused on as an important issue in terms of telelearning and in terms of adaptive public policies (which feature later in the book). Observations were also made that: most knowledge of innovation comes from a network of sources; institutions are typically more adept at utilizing knowledge from inside the organization than from outside; training is not readily isolated as part of an organization's strategy; and management styles tend to be as important as management structures. In addition, despite the widespread discussion of high skill requirements and high trust work practices, comparative studies show that such firm level best practices are not being readily diffused. Thus again we learn of the importance of social capital and networks. Policy and strategy implications suggest that human capital is a much needed focus of attention.

3. MEASURING OBSERVATIONS

Authors highlighted here a number of issues: internet dropout rates, business-to-household issues in information interaction, and the relationship between ICT development and use and its impacts on social and economic patterns. Some of the chapters reported on the first Canadian efforts to measure the pervasiveness of Internet technology and serve as useful benchmarks. However, as the authors made clear, much remains to be done, in the fashion outlined above. In a very useful paper, Georg Licht dealt in part with the relationship between public research institutions and SMEs. This is important given the valuable role of technical assistance that federal labs and research infrastructures can play in countries like Canada and Germany. Indeed, many service companies possess excellent preconditions for successful cooperation with public research institutions. The high educational level of employees, particularly in research and technical areas, often can serve to lower barriers to cooperation. Measurement observations such as these have myriad policy and strategy lessons.

4. IMPACT OBSERVATIONS

As the collection of chapters in this section highlighted, the impact of information and ICTs on social, economic, cultural, political and legal systems are, at times, head-spinning. In his discussion of the global corporation, Cliff Wymbs usefully highlights the information value chain as a means to re-defining the activities and parameters of the firm. Valerie Steeves argues articulately that the link between privacy and private property is not random. Architecture, at least in the legal sense, *is* policy. In this brief motto, she focuses valuably on the need to continue to keep our legal and technological choices squarely in mind as we morph into a networked society. Laura Winer, balances key operational questions about access, cost, timing, and quality of training/learning with social (one might say policy) questions concerning equality and inclusion. Interestingly, few of Winer's issues on telelearning are strictly educational. But, as she says, "the fundamental goal of education and training is to support individuals – *and by extension groups, communities, organizations, and our society* – to survive in, adapt to, and ultimately influence for the better, our world" (emphasis added). This indeed is an onus evident throughout the section and the volume. Extending the concern with learning and education is Nico Stehr's exploration of the world of the knowledge worker, the labour market and the productivity paradox. "As long as economic organizations and regimes such as the labour market are reconstructed as inflexible and largely one-dimensional, the gradients of inequality will be steep and the effects of its logic long lasting…. [New] realities require a new language of the world of work. The new language should…. stress agency, malleability, flexibility, multi-purpose resources, volatility, heterogeneity, etc. of social structures". One aspect of this is that socio-economic adjustment will not lay all responsibility at the feet of workers but at the feet of managers, policy makers and institutions as well. John de la Mothe extends many of the foregoing conclusions into the realm of public management and policy. He reminds that the seemingly new public debate concerning the knowledge-based society is historically grounded and embodies implicit reactions towards technology and government. He highlights how this influences the contemporary impacts of information and ICTs from political, public management and public policy perspectives, and goes on to call for the need for adaptive public approaches in a networked socio-economy.

But as valuable as these collective observations are, where do they lead us? Some directions are suggestive.

5. FUTURE DIRECTIONS

5.1 Statisticians

For official statisticians, there is an obligation to provide information on the production of information and communication technologies and on the data, information, and knowledge products that ICTs make pervasive. Such economic statistics should encompass all of the components of Gross Domestic Product, as well as labour and trade statistics. For this to happen, there has to be consensus on what industries belong to the ICT sector, and on the products produced by the sector. The OECD has provided an industry-based definition of the former and is working on the latter. However, the real challenge for the official statistician is a description of the 'content' sector, and a classification of content products which in this context means electronic content products which are produced by some industries for the electronic display, processing, storage and transmission by ICT goods and services. Little progress on the qualitative analysis of the information economy or the information society will be made until these issues of classification are resolved.

Providing a statistical description of the production system for ICTs and for content is but a first step, albeit one of fundamental importance. The issue of importance to decision makers in firms and government, and to people, is the impact of the outcomes of these production activities. As an example, consider the effect the cellular telephone is having on industrial organization and on the learning practices of people. Before such analyses can be performed for ICT and 'content' products, there must be agreement on what they are. This, again, is a classification issue that official statisticians must resolve.

As several papers in this volume have noted, the activities of innovation, production, or use do not take place in isolation. The activity of innovation is linked to the economic context. This includes government regulation and incentive programs, the knowledge in government and university laboratories as well as the market behavior and influences of suppliers and customers. Innovation also has outcomes that may include higher employment levels in successful firms of more highly skilled people. The outcomes may not be so positive. In either case there is a need, recognized by Statistics Canada and noted by several authors, to measure – not just innovation or R&D but also – the linkages of the actors engaged in the activities. This need must be addressed before moving on to the even more difficult challenge of measuring the economic and social impacts of such activities.

These issues leave official statisticians with a very full agenda. However this work also has significant private and public policy implications.

5.2 Policy Makers and Analysts

For corporate managers, the processes of competition and innovation – and the pressures arising from their interaction with public, social and cultural institutions and trends - will jointly continue to press forward and shape the development of ICT products and processes. This in itself will be highly challenging. But for policy makers and public managers, there is a more complex series of tasks.

As with the statisticians' challenges of clearing the classification underbrush, so too must policy makers and analysts not get caught up with a techno-enthusiasm in which advancing an industrial-trade agenda is their sole raison d'être and the mechanism for trickle down benefits. Indeed, policy considerations, as the book makes abundantly clear, must attend to a multiplex of simultaneous issues. So, one job is to disaggregate.

At one level, ICTs have made possible the broad re-engagement of government and politics to the concerns of its citizens. The issue here is not so much that technologies have made politicians and public servants potentially more accessible, or that government information can be made more readily available to communities or non-traditional users. Instead the issue pivots on the degree to which the institutions and agencies of government are able to be responsive to this emerging channel of two-way, group-way or mega-way communication between government officials and their constituents, and to manage - and (importantly) to evaluate – the flows and quality of information. This will determine whether the hierarchical arrangements that have typified bureaucracy for so long can be made more porous and effective - facilitating horizontal coordination across agencies, transparency, networked decision making, and accountability. Such moves will be essential if governments are to maintain political authority domestically in an era of increased international integration as well as to guide their own socio-economies in the global, de-materialized context.

Additionally, the information society entails not only a different set of rules for the new economy but also demands a different set of tools with which to manage the new realities. Public policies and programs today help govern, stimulate and shape <u>spaces</u> as well as <u>places</u>. Thus, a simultaneous blend of unevenly developed regions in a country – representing the spectrum from 'smart cities' and 'backward regions' will place strains on public decision making for equitable treatment. In order to avoid reaching for the lowest common dominator in which nothing is common, it will rely on the ability of civil servants - in consultation with interested parties (internal to the government as well as external), constituents and other relevant levels of government – to creatively design programs which are uniform in architecture but flexible and responsive to the requirements of a specific region or locale. Extant examples of ICTs being used to provide better government services or to custom design programs include: employment training programs that might focus on the energy sector in Texas or Alberta and the software industry in Saint John or Brighton; the ability to file taxes electronically; the ability to file for unemployment insurance, health or welfare benefits in shopping malls, and so on.

These broad observations require public management to decipher and clearly focus on the myriad impacts and implications of ICTs and information on the re-shaping of their work.

Thus, public policy in an effective Information Society will need to focus expressly on the fostering and animation of social capital. This can be accomplished from a variety of government departmental purviews – such as those responsible for industry and commerce, trade, investment, culture and heritage, foreign affairs and environment. This of course does not require choosing between technology programs *or* content initiatives. Instead, it requires a deliberate coupling of technology-socio-cultural themes. An example might be the strategic decision to pursue multi-media educational software as a national goal which then demands a blend of technical and social dimensions. Moreover, it would also require a deliberate attempt to horizontally coordinate within government and might profitably gain from transversal coordination between communities.

Such a move would lead to the facilitation of another series of needed policy orientations. One is to focus on incentives and methods to facilitate the diffusion and adoption of ICTs in a variety of industrial sectors, as well as in non-governmental sectors including schools, libraries, and so on. Another is to adopt the principles of networks or innovation systems[1] and devolve decision making and decision inputs down to communities and interest pockets. By focusing on social inclusion, not only will the potential distortion of one group dominance be avoided but the needed initiatives towards social capital will featured.

At an almost regulatory level, an important role for policy makers will be in the domain of standards and intellectual property. This will require the striking of an aggressively proactive three-sided balancing act between: (a) ensuring the macro-economic stability and the investment allure of an economy such that innovative behaviour in firms and labs can be encouraged and can participate in the advance of world technological developments; (b) helping to set and harmonize world terms and conditions for IP protection, 'rules of the game', etc.; and (c) ensuring the protection of indiviudal information. Thus, as was suggested above, the challenges to policy makers are considerable. And in part they will require the active engagement of the research communities.

5.3 Research Communities

Research communities are critical not only in their allotted function as critical analysts. They also play an essential role of providing a feedback mechanism between statisticians, policy makers and the public. Here, researchers can offer statisticians valuable reflections on trends, methodologies, linking data sets, and so on. Similarly, they can provide policy makers not only with much needed information regarding statistical and research insights, but they can also reveal gaps and impacts in government programs and policy frameworks. Since academic researchers tend to be more regionally distributed that public managers, this added coverage can be invaluable in facilitating a conduit between stakeholders and acting as an agent for flexibility, responsiveness and reaction.

More specifically, in an Information Society, there is much that is not understood about the dynamics of innovation, networking, coordination, and flexibility. There is much that is not understood about the nexus between social systems and technical systems. There is much that is not yet appreciated about the mutual impacts of ICTs on social, cultural and economic systems. Researchers – particularly in the social fields of economics, sociology, anthropology, management, history and statistics as well as in some of the arts – have a difficult but urgent task to aggressively embrace these issues, *not* in the siloed isolation of their 'fields'. Each of these questions is by definition discipline defiant. As such, the research community needs to also embrace the principles outlined in the network society. Partnerships, across sectors and between fields, will – as much as the policy makers and the statisticians have been advised – yield constructive (and no doubt at times controversial) observations which will permit the advancement of ICTs while promoting social inclusion, political cohesion and cultural vitality.

Thus, in a world where measurement, policy and research interact, instantaneously and non-linearly, there is no isolation. The issue for all of the actors in the system is interaction, across social and economic boundaries. As Karl Popper once put it, "in the beginning, always, is the issue". Interaction is the issue that begins this volume, and that ends it.

Notes

[1] This has been discussed in a previous Statistics Canada – PRIME workshop and book. See John de la Mothe and Gilles Paquet (eds.), 1998, *Local and Regional Systems of Innovation*, Boston, Kluwer.

References

Adam, N. & Yesha, Y., (1997). "Electronic Commerce and Digital Libraries: Toward a Digital Agora", *ACM*, February.

Adam, N., Awerbuch, S., Slonim, J., Wegner, P. & Yesha, Y., (1996). "Globalization: Technology and Social Impact", *ACM*, December.

Adams, E., (1994). "Better Connected", *World Trade*, Vol. 7, No. 8, Pp. 98-100, September.

Aghion, P. & Howitt, P., (1998). *Endogenous Growth Theory*, Cambridge, Massachusetts, MIT-Press.

Agre, P.E. & Rotenberg, M., (1997). *Technology and Privacy: The New Landscape*, Cambridge MA, MIT Press.

Akerlof, G.A. and W.T. Dickens, (1982). "The Economic Consequences of Cognitive Dissonance", *American Economic Review* 72 (3), 307-319

Alexander, C. (1964). *Notes Toward a Synthesis of Form*. Cambridge, Harvard University Press.

Alexander, T.J., (1997). "Human Capital Investment: Building the 'Knowledge Economy", *Policy Options*, Vol. 18, Pp. 5-8.

Allen, L. E., (1997). "Home, Home, on the Web", *Mortgage Banking*, Vol. 57, No. 6, Pp. 46-54, March.

Ang, P. & Nadarjan, B., (1996). "Censorship and the Internet: A Singapore Perspective", *Communications of the ACM*, Vol. 39, No. 3, June.

Antonelli, C., (1996). "Localized Knowledge Percolation Processes and Information Networks", *Journal of Evolutionary Economics* 6 (3), 281-295.

Antonelli, C., (1998). Localized technological change, new information technology and the knowledge-based economy: the European evidence. *Journal of Evolutionary Economics* 8 (2) 177-198.

Argyris, C., (1977). "Double Loop Learning in Organizations", *Harvard Business Review*, Sept.-Oct.

Argyris, C. & Schon, D., (1978). *Organizational Learning: A Theory of Action Perspective*, Reading, Massachusetts, Addison-Wesley.

Aristotle, (1970). "Knowledge and Wisdom", from Book IV of the Oxford edition of *Nichomachean Ethics*, translated by Ross, W.D., Clarendon Press, Oxford.

Ark, B., van Monnikhof, E. & Mulder, N., (1999). "Productivity in Services: An International Comparative Perspective", *Canadian Journal of Economics*, Vol. 32, Pp. 471-499.

ASTD, (1998). "National HRD Executive Survey Information Technology Training 1998 Second Quarter Survey Report", http://www.astd.org/virtual_community/research/nhrd_executive_survey_98it.html.

Attewell, P. and Battle, J., (1999). Home computers and school performance. *The Information Society* 15 (1), 1-10.

Attewell, P., (1994). "Information Technology and the Productivity Paradox", in Harris, D.H. (Ed.), *Organizational Linkages, Understanding the Productivity Paradox*, Pp. 13-53, Washington, D.C., National Academy Press.

Baldwin, J. & Johnson, J., (1995). *Human Capital Development and Innovation*, Ottawa, Statistics Canada.

Baldwin, J., (1995). *The Dynamics of Industrial Competition – A North American Perspective.* Cambridge: Cambridge University Press.

Baldwin, J.R. & Gellatly, G., (1998). *Are There High-Tech Industries or Only High-Tech Firms? Evidence from New Technology-Based Firms*, Ottawa, Statistics Canada, Analytical Studies Branch.

Baldwin, J., Gellatly, G., Johnson, J. & Peters, V., (1998). "Innovation in Dynamic Service Industries", Statistics Canada, Ottawa, Catalogue No. 88-516-XIE.

Baldwin, J.R. & Rafiquzzaman, M., (1998). "Trade, Technology, and Wage Differentials in the Canadian Manufacturing Sector", in Audretsch, D.B. & Thurik, A.R. (Eds.), *Innovation, Industry Evolution, and Employment*, Cambridge, Cambridge University Press.

Barney, J., (1991). "Firm Resources and Sustainable Competitive Advantage", *Journal of Management*, Vol. 17, Pp. 99-120.

Barras, R., (1986). "Towards a Theory of Innovation in Services", *Research Policy*, Vol. 15, No. 4.

Barras, R., (1990). "Interactive Innovation In Financial And Business Services: the Vanguard of the Service Revolution", *Research Policy*, Vol. 19.

Bassi, L., (1997). "Harnessing the Power of Intellectual Capital", ASTD, Alexandria, Virgina.

Bassi, L. & Van Buren, M. E., (1998). *State of the Industry Report*, Alexandria, Virginia, American Society for Training & Development.

Bassi, L. & Van Buren, M., (1998). "Investments in Intellectual Capital: Creating Methods", 2nd World Congress on Management of Intellectual Capital.

Batchelor, R. A. & Orr, A. B., (1988). "Inflation Expectations Revisited", *Economica*, Vol. 55, Pp. 317-331.

Baumard, P. (1996). *Organisations déconcertées: la gestion stratégique de la connaissance.* Paris, Masson.

Bay, A., (1995). "Keynote Address", Internet@Telecom 95, October 8.

Belanger, M., (1996). "Canada's InfoWay The So-Called "Information Highway", http://www.solinet.org/library.html.

Bell, D., (1968). "The Measurement of Knowledge and Technology", in Sheldon, E.B. & Moore, W.E. (Eds.), *Indicators of Social Change. Concepts and Measurements*, Pp. 145-246, Hartford, Connecticut, Russell Sage Foundation.

Bell, D., (1973). *The Coming Post-Industrial Revolution*, New York, Basic Books.

Belleflamme, C., Houard J. & Michaux, B., (1986). *Innovation and Research and Development Process Analysis in Service Activities*, Brussels, EC, FAST Occasional papers No. 116.

Berman, E., Bound, J. & Geliches, Z., (1994). "Changes in the Demand for Skilled Labor within U.S. Manufacturing Industries: Evidence from the Annual Survey of Manufacturing", *Quarterly Journal of Economics*, Vol. 59, Pp. 367-398.

Bernard, A.B. & Jensen, J.B., (1997). "Exporters, Skill Upgrading and the Wage Gap", *Journal of International Economics*, Vol. 42, Pp. 3-31.

Berry, J.M., (1999). *The New Liberalism: The Rising Power of Citizen Groups*, Washington, Brookings.

Bessant, J. & Rush, H., (1998). "Building Bridges for Innovation: the Role of Consultants in Technology Transfer", *Research Policy*, Vol. 24.

Best, M. (1990). *The New Competition*. Cambridge: Harvard University Press.

Betcherman, G. et. al., (1998). *Training for the New Economy*, Ottawa, Canadian Policy Research Networks

Bijker, W.E., & Law, J. (Eds.), (1992). *Shaping Technology – Building Society*, Cambridge, MIT Press.

Bijker, W.E., Hughes, T.P. & Pinch, T. (Eds.), (1987). *The Social Construction of Technological Systems: New Directions in the Sociology and History of Technology*, Cambridge, Massachusetts, MIT Press.

Bilderbeek, R. & den Hertog, P., (2000, in press). "The New Knowledge Infrastructure: The Role of Technology-Based Knowledge-Intensive Business Services in National Innovation Systems", in Miles, I. & Boden, M. (Eds.), *Knowledge, Innovation and Services*, London, Pinter.

Black G., (1997). "The Promise of High Returns", *Financial Times*, April 2, 1997. P. 10.

Bledsoe, C., (1994). "Constructing Natural Fertility: The Use of Western Contraceptive Technologies in The Gambia", *Population and Development Review*, Vol. 20, No. 1, Pp. 81-113.

Blundon, W., (1996). *Internet World*, Vol. 7, No. 12, http://www.iworld.com.

Böhle, F., (1998). "Technik und Arbeit – neue Antworten auf 'alte' Fragen", *Soziale Welt*, Vol. 49, Pp. 233-252.

Boisot, M. (1995). *Information Space — A Framework for Learning in Organizations, Institutions and Culture*. London, Routledge.

Bommer, M. & Lalajas, D. S., (1999). "The Threat of Organizational Downsizing on the Innovative Propensity of R&D Professionals", *R&D Management,* Vol. 29, No. 1, Pp. 27-34.

Bontis, N., (1996). *Intellectual Capital: An Exploratory Study that Develops Measures and Models*, "17th McMaster Business Conference" Paper.

Bourdieu, P., (1973). "Cultural Reproduction and Social Reproduction", in Brown, R., (Ed.), *Knowledge, Education, and Cultural Change*, Pp. 71-112,

Bracewell, R., Breuleux, A., Laferrière, T., Benoit, J. & Abdous, M., (1998). "The Emerging Contribution of Online Resources and Tools to Classroom Learning and Teaching", report submitted to SchoolNet/Rescol by Telelearning Network Inc., http://www.telelearn.ca/g_access/news/review.html.

Branscomb, L., *et. al.*, (1994). *Empowering Technology*, Camrbridge, MIT Press.

Bransten, L., (1997). "Shopping on the Internet", *Financial Times*, April 1, 1997. P. 10.

Braudel, F., (1967). *Civilisation materielle et capitalisme*, XV e - XVII e siecle, Paris, Armand Colin.

Braverman, H., (1974). *Labor and Monopoly Capital: The Degradation of Work in the Twentieth Century*, New York, Monthly Review Press.

Briggs Myers, I., (1993). *Introduction to Type*, Flint, Consulting Psychologists Press Inc.

Brin, D., (1999). *The Transparent Society*, New York, Addison Wesley.

Brooking, A., (1996). *Intellectual Capital: Core asset for the third millennium enterprise*, London, International Thomson Business Press.

Bruinshoofd, A. and Ter Weel, B., (1998). *Skill-biased technical change: on technology and wages in the Netherlands.* Maastricht Economic Research Institute on Innovation and Technology (MERIT), Research Memorandum 98-025.

Brynjolfsson, E. & Hitt, L., (1996). "Paradox Lost? Firm-Level Evidence on the Returns to Information Systems Spending", *Management Science*, Vol. 42, Pp. 541-558.

Burge, E.J. & Roberts, J.M., (1998). *Classrooms with a Difference: Facilitating Learning on the Information Highway*, (Second Edition), Montréal, Chenelière/McGraw-Hill.

Burgoyne, J. et. al., (1994). *Towards the Learning Company*, London, McGraw-Hill, Butterworth-Heinemann.

California Management Review, (1998) Special issue on Knowledge and the Firm, Spring.

Canada, (1996). "Building the Information Society: Moving Canada into the 21st Century", Ottawa.

Canada, (1996). "Measuring the Global Information Infrastructure for a Global Information Society: Concepts and Performance Indicators", document submitted to the OECD, September.

Carlson, J.A. & Parkin, M. M., (1975), "Inflation Expectations", *Economica*, Vol. 42, Pp. 123-138.

Castells, M., (1989). *The Informational City: Information Technology, Economic Restructuring, and the Urban-Regional Process*, Oxford, Basil-Blackwell.

Castells, M., (1996). *The Information Age: Economy, Society and Culture, Volume 1, The Rise of the Network Society*, Malden, Massachusetts, and Oxford, United Kingdom, Basil Blackwell.

Cavestro, W., (1989). "Automation, New Technology and Work Content", in Wood, S. (Ed.), *The Transformation of Work?*, Pp. 219-234, London, Unwin Hyman.

CBI/Natwest, (1996). "Innovation Trends Survey", London, Natwest Innovation and Growth Unit.

Chase, R., (1997). "The Knowledge-Based Organization: An International Survey", *The Journal of Knowledge Management*, Vol. 1, No. 1, September, Pp. 38-49.

Chizzo, S. A., (1998). "Supply Chain Strategies: Solutions for the Customer-Driven Enterprise", *Software Magazine*, Pp. 4-9. January.

Chodorowicz, D. & Sciadas, G., (1998). "The Cellular Telephone Industry: Birth, Evolution, and Prospects", *Canadian Economic Observer*, Cat. No. 11-010-XPB, August.

Choi, J.P., (1997). Herd Behaviour, the "Penguin Effect," and the Suppression of Informational Diffusion: an Analysis of Informational Externalities and Payoff Interdependency. *RAND Journal of Economics* 28 (3) 407-425.

Chudleigh, M. & Sheppard, M., (1995). "Electronic Commerce: Revenue Canada Helping Canadian Companies Save Time and Money". *Cost and Management*, Vol. 69, No. 5, June.

Cimoli, M. & de la Mothe, J., (2000). "Technology, Growth and Development: Falling Behind, Catching Up, and 'Leap Frogging' Ahead", in de la Mothe, J. (Ed.), *Science, Technology and Governance*, London, Pinter.

Clinton, W.D., (1994). *The Two Faces of National Interest*, Baton Rouge, Louisiana State University Press.

CMA, (1998). *The Management of Intellectual Capital: The issues and the practice*, Society of Management Accountants of Canada, Hamilton, Ontario.

Coase, R.H., (1937). "The Nature of the Firm", *Economica*, Vol. 4 (November), 386-405.

Colecchia, A., (1999). Presentation at the Workshop on "Statistical Challenges of Connectedness and Electronic Commerce", Ottawa, February 18-19.

Coleman, J., (1988). "Social Capital in the Creation of Human Capital," *American Journal of Sociology Supplement*, 94, Pp. S95-S120.

Communication From The Commission To The Council, (1998). *The Contribution of Business Services to Industrial Performance: A Common Policy Framework*, Brussels, 21.09.1998 COM (1998) 534 final.

Conference Board, (1997). "Leveraging Intellectual Capital", *HR Executive Review*, Vol. 5, No. 3.

Coward, L.A. (1990). *Pattern Thinking*. New York, Praeger.

CRTC, (1999). *CRTC Won't Regulate the Internet*, News Release, May 17, 1999, http://www.crtc.gc.ca/ENG/NEWS/RELEASES/1999/R990517e.htm.

Curry, J., Neill, W.R., Galan, N. & Harasim, L., (1998). *The Telelearning Industry: An Overview of Activities and Opportunities in North America*, http://www.telelearn.ca/g _access/news/TLindustry.html.

Dankbaar, B., (1995). "The Crisis of Fordism: Restructuring in the Automobile Industry", in van Ruysseveldt, R.H. & van Hoff, J. (Eds.), *Comparative Industrial and Employment Relations*, Pp. 293-314, London, Sage.

Darlington, L., (1998). "Banking without Boundaries", in Tapscott, D., Lowy, A., & Ticoll, D. (Eds.), *Blueprint to the Digital Economy: Creating Wealth in the Era of E-business*, New York, N.Y., McGraw Hill.

Das, S., (1999). "Knowledge Seen as the Base of Future Society," *The Edmonton Journal*, 10 February 1999, P. F3.

Davenport, P., (1997). "The Productivity Paradox and the Management of Information Technology", Paper presented to the Centre for the Study of Living Standards conference on "Service Sector Productivity and the Productivity Paradox", April 11-13, 1997, Ottawa, Canada.

Davenport, T.H. & Prusak, L., (1998). *Working Knowledge*, Boston, Harvard Business School Press.

David, P., (1990). "The Dynamo and the Computer. An Historical Perspective on the Modern Productivity Paradox", *American Economic Review*, Vol. 80, Pp. 355-361.

Davis, J., Hirschl, T. & Stack, M. (Eds.), (1997). *Cutting Edge: Information, Capitalism and Social Revolution*, New York, Verso.

de Jong, M. W., (1994). "Core Competencies and Chain Relations in Service Industries", in Institut d'Administration des Enterprises, *Management of Services: A Multidisciplinary Approach*, Proceedings of 3rd International Research Seminar in Service Management, Marseille, Universite d'Aix-Marseille III.

De Jonquieres, G., (1997). "Internet: Holy Grail or Fool's Goal", *Financial Times*, February 4, P. 8.

de la Mothe, J., (1989). "Science and Literature in the Age of the Cyber", *Canadian Journal of Political and Social Theory*. Vol. 13, No.1/2, Pp.167-175

de la Mothe, J., (1999). "Government Science and the Public Interest", Science ADM's Committee, Ottawa, March.

de la Mothe, J. & Ducharme, L.M., (1990). *Science, Technology and Free Trade*, London, Pinter.

de la Mothe, J. & Dufour, P., (1995). "Techno-Globalism and the Challenges of Science and Technology Policy", *Daedalus: The Journal of the American Academy of Arts and Science*, Vol. 124, No. 3.

de la Mothe, J. & Paquet, G., (1992). "Local and Regional Systems of Innovation as Learning Socio-Economies", in de la Mothe, J. & Paquet, G. (Eds.), *Local and Regional Systems of Innovation*, Norwell, Kluwer.

de la Mothe, J. & Paquet, G., (1994a). "The Dispersive Revolution", *Optimum: The Journal of Public Sector Management*, Ottawa.

de la Mothe, J. & Paquet, G., (1994b). "The Technology-Trade Nexus", *Technology in Society*, Vol. 16.

de la Mothe, J. & Paquet, G. (Eds.), (1996a). *Evolutionary Economics and the New International Political Economy*, London, Pinter.

de la Mothe, J. & Paquet, G., (1996b). "Circumstantial Evidence: A Note on Science and Technology Policy in Canada", *Science and Public Policy*.

de la Mothe, J. & Paquet, G., (Eds), (1997). *Challenges Unmet in the New Production of Knowledge*. Program of Research on International Management and Economy, University of Ottawa.

de la Mothe, J. & Paquet, G., (1998a). "Finance and the Technology-Trade Nexus", *Technology in Society*, Vol. 20.

de la Mothe, J. & Paquet, G. (Eds.), (1998b). *Local and Regional Systems of Innovation*, Norwell, Kluwer.

den Hertog, P. et al., (1997). "Intangibles: The Soft Side of Innovation", *Futures*, Vol. 29, No. 1.

Denton, N., (1997). "Drive to Plug the Gap", *Financial Times*, February 3, P. 19.

Dickinson, P. & Ellison, J., (1999). "Getting Connected or Staying Unplugged: The Growing Use of Computer Communications Services", *Services Indicators*, Vol. 6, 1, Statistics Canada, Catalogue 63-016-XPB, Ottawa.

Dickinson, P. & Sciadas, G., (1996). "Access to the Information Highway", *Canadian Economic Observer*, Statistics Canada, Catalogue No. 11-010-XPB, December.

Dickinson, P. & Sciadas, G., (1997). "Access to the Information Highway: the Sequel", *Services Indicators*, Statistics Canada, Catalogue No. 63-016-XPB, June.

Dickinson, P. & Sciadas, G., (1999). "Canadians Connected", Canadian *Economic Observer*, Statistics Canada, Catalogue No. 63-016-XPB, February, 3.1-3.22.

Dierkes, M., Hofman, J. & Mars, L., (1998). "Technological Development and Organisational Change: Differing Patterns of Innovation", in, *21st Century Technologies: Promises and Perils of a Dynamic Future*, Pp. 97-122, OECD, Paris.

Diewert, E. & Fox, K., (1997). "Can Measurement Error Explain the Productivity Paradox?", Paper presented to the Centre for the Study of Living Standards conference on "Service Sector Productivity and the Productivity Paradox", April 11-13, Ottawa, Canada.

Diewert, W.E. & Fox, K.J., (1999). "Can Measurement Error Explain the Productivity Paradox?" *Canadian Journal of Economics*, Vol. 32, No. 2, Pp. 251-280.

DIW, (1998). "Das Dienstleistungs-Puzzle. Ein aktualisierter deutsch-amerikanischer Vergleich", DIW-Wochenbericht 35/98, Pp. 625-629.

Dixon, N., (1993). *Organizational Learning*, Conference Board of Canada.

Doern, G.B., (1998). *Intellectual Property Organizations and Global Change*, London, Pinter.

Doern, G.B., Pal, L.A. & Tomlin, B.W. (Eds.), (1996). *Border Crossings: The Internationalization of Canadian Public Policy*, Toronto, Oxford University Press.

Dohmen, D., (1999). *Ausbildungskosten, Ausbildungsförderung und Familienlastenausgleich – Eine ökonomische Analyse unter Berücksichtigung rechtlicher Rahmenbedingungen*, Berlin, Duncker und Humblot.

Doms, M., Dunne, T. & Troske, K., (1997). "Workers, Wages and Technology", *The Quarterly Journal of Economics*, February, Pp. 253-290, Doubleday.

Dosi, G. 1996. The Contribution of Economic Theory to the Understanding of a Knowledge-Based Economy, in *Employment and Growth in the Knowledge-Based Economy*. Paris, OECD, 81-92.

Downes, L. & Mui, C., (1998). *Unleashing the Killer App: Digital Strategies for Market Dominance*, Cambridge, Mass., Harvard Business School Press.

Doyle, R., (1997). "By The Numbers: Access to the Internet", *Scientific American*, Vol. 227, No. 1, P. 26.

Drucker, P., [1968] (1992). *The Age of Discontinuity. Guidelines to Our Changing Society*, (With a New Introduction by the Author), New Brunswick, New Jersey, Transaction Books.

Drucker, P., (1999). *Management Challenges for the 21st Century*, New York, Harper Business.

Dudley, L., (1996). Comme un canari en cage: leçons a tirer de l'effondrement de la RDA. In: Coulombe, S., and Paquet, G. (Eds) *La réinvention des institution et le role de l'État*. Montreal, ASDEQ, 379-392.

Dumort, A. & Dryden, J. (Eds.), (1997). "The Economics of the Information Society", European Commission and the OECD.

Dunning, J. H. & Wymbs, C., (1997). "The Geographical Sourcing of Technology Based Assets by Multinational Enterprises," Newark, New Jersey, Rutgers University Mimeo.

Dunning, J. H., (1988). *Explaining International Production*, London & Boston, Unwin Hyman.

Dunning, J. H., (1993). *Multinational Enterprises and the Global Economy*, New York, Addison-Wesley.

Dunning, J. H., (1995). "Reappraising the Eclectic Paradigm in an Age of Alliance Capitalism", *Journal of International Business*, Third Quarter, Pp. 461-491.

Easlea, B., (1984). *Liberation and the Aims of Science*, Brighton, Belhaven.

Edvinsson, L. & Malone, M.S., (1997). *Intellectual Capital: Realizing Your Company's True Value by Finding its Hidden Brainpower*, New York, Harper Business.

Eiglier, P. & Langeard, E., (1987). *Servuction*, Paris, McGraw-Hill.

Eika, K.H. and Reistadbakk, T., (1998). The competitive environment in the banking industry – driving forces and trends. *Norges Bank Economic Bulletin* 69 (4), 368-380.

Ekos Research Assoc. & Lyndsay Green & Assoc., (1999). *The Impact of Learning Technologies on Learning in the Workplace*, http://olt-bta.hrdc-drhc.gc.ca/publicat/ index.html#ekos.

Elster, J. (1989). *Nuts and Bolts for the Social Sciences.* Cambridge, Cambridge University Press.

Engler, N., (1997). "The Second Coming of Electronic Commerce", *Computerworld*, Vol. 31, No. 50, Pp. E10-16, Dec 15.

European Commission, (1997). *Building the European Information Society for Us All*, Luxembourg, April.

European Foundation For The Improvement Of Living And Working Conditions, (1997). *New Forms of Work Organisation. Can Europe Realise its Potential?*, Office for Official Publications of the European Communities, Luxembourg.

Evans, P. B. & Wurster, T.S., (1997). "Strategy and the New Economics of Information", *Harvard Business Review*, Vol. 75, No. 5, September/October, Pp. 70-82.

Ferrand, D. and Paquet, G. (1994). Apprentissage organisationnel et ré-engineering. *Logistique et Management* 2(1), 45-58.

Fleming, M., (1997). "The New Business Cycle: the Impact of the Application and Production of Information Technology on U.S. Macro-economic Stabilization". *Business Economics* 32 (4), 36-42.

Florida, R. (1991). The New Industrial Revolution. *Futures* 23(6), 559-576.

Florida, R., and Kenney, M. (1993). Innovation-Mediated Production. *Futures* 25(5), 637-651.

Foray, D. & Lundvall. B.A., (1996). "The Knowledge-Based Economy: From the Economics of Knowledge to the Learning Economy", in *Employment and Growth in the Knowledge-Based Economy*, Paris, OECD, 11-32.

Forbes magazine, (1997). Special Section: "Measurement", Pp. 35-82, April 17.

Foster, J. and Wild, P. (1996). Economic Evolution and the Science of Synergetics. *Journal of Evolutionary Economics*. 6 (3), 239-260.

Freeman, C., (1996). "Catching Up and Falling Behind: The Case of Asia and Latin America", in de la Mothe, J. & Paquet, G. (Eds.), *Evolutionary Economics and the New International Political Economy*, London, Pinter.

Freeman, C. & Soete, L., (1994). *Work for All or Mass Unemployment?: Computerised Technical Change into the 21st Century*, London, Pinter.

Freeman, C., Soete, L., & Efendioglu, U., (1995). Diffusion and the employment effects of information and communication technology. *International Labour Review* 134 (4-5), 587-603.

Fricke, W., (1997). "Evaluation of the German Work and Technology Programme from an Action Research Point of View", in Alasoini, T., Kyllönen, M. & Kasvio, A. (Eds.), *Workplace Innovations – A Way of Promoting Competitiveness, Welfare and Employment*, National Workplace Development Programme, Yearbook 1997, Helsinki.

Fujimoto, T., (1998). "A Case of the Capability Evolution: Toyota and Japanese Auto Industry", *La lettre du Gerpisa*, No. 119, January.

Fukuyama, F., (1992). *The End of History and the Last Man*, New York, Free Press.

Fukuyama, F., (1995). *Trust: Social Virtues and the Creation of Prosperity*, New York, Free Press.

Fukuyama, F., (1999). "The Great Disruption", *The Atlantic Monthly*, May, New York, Free Press.

Galbraith, J.K., (1967). *The New Industrial State*, Boston, Houghton Mifflin.

Gallagher, A.L., (1997). *Intellectual Capital: A Summary of Corporate Initiatives*, Alexandria, Virginia, American Society for Training & Development.

Gandell, T., Weston, C., Finkelstein, A. & Winer, L.R., (in preparation). *Using the Web in Teaching in Higher Education*, Montreal, CUTL, McGill University.

Gärdin, O., (1997). "Statistics for the Information Society and the Networked Economy", in Dumort, A. & Dryden, J., *The Economics of the Information Society*, European Commission and the OECD.

Garton, L., Haythornthwaite, C. & Wellman, B., (1999). "Studying On-Line Social Networks", in Jones, S. (Ed.), *Doing Internet Research*, Thousand Oaks, CA, Sage, P. 75-105.

Garvin, D.A., (1993). "Building a Learning Organization", *Harvard Business Review*,

Gault, F., (1997). "Canadian Statistics on the Information Society", in Dumort, A. & Dryden, J., *The Economics of the Information Society*, European Commission and the OECD.

Gephart, M. & van Buren, M., (1996). "High Performance Work Systems", Washington, D.C., ASTD Workshop.

Gera, S. & Philippe, M., (1996). *Employment Performance in the Knowledge-Based Economy*, Ottawa, Industry Canada/HRDC Working Paper.

Gera, S., Sing, C.L. & Newton, K., (1998). *The Emerging Global Knowledge Based Economy*, paper prepared for the Industry Canada/CYRANO conference,

Ghosh, S., (1998). "Making Business Sense of the Internet", *Harvard Business Review*, March/April, Pp.126-135.

Giarini, O. & Stahel, W. R., (1993). *The Limits to Certainty, Facing Risks in the New Service Economy*, 2nd ed., Dordrecht, Boston, London, Kluwer Academic Publishers.

Gibbons, M. (1994). Transfer Sciences: Management of Distributed Knowledge Production. *Empirica*, 21, 259-270.

Gibbons, M. et al., (1994). *The New Production of Knowledge*, London, Sage.

Giddens, A., (1994). *Beyond Left and Right*, Cambridge, Polity Press.

Giddens, A., (1999). *The Third Way*, Cambridge, Polity Press.

Gill, C., (1985). *Work, Unemployment and the New Technology*, Oxford, Polity Press.

Gilles, W. and Paquet, G. (1991). La connaissance de type Delta, in G. Paquet and O. Gélinier, Eds, *Le management en crise: pour une formation proche de l'action*. Paris, Economica, 19-36.

Goldbert, M.W., (1997). "Communication and Collaboration Tools in WebCT", proceedings of the conference, "Enabling Network-Based Learning", May 28-30, 1997, Esposo, Finland, http://www.webct.com/webct/papers/.

Gore, A., (1994). "Creating a Government that Works Better and Costs Less: From Red Tape to Results", Washington, Washington Printing Office.

Gortmaker, S.L. & Wise, P.W., (1997). "The First Injustice: Socioeconomic Disparities, Health Services Technology and Infant Mortality," *Annual Review of Sociology*, Vol. 23, Pp. 147-170.

Gottschalk, P., (1997). "Inequality, Income Growth and Mobility", *Journal of Economic Perspectives*, Vol. 11, Pp. 21-40.

Government of the U.K., (1996). "Development of the Information Society: An International Analysis", a report by Spectrum Strategy Consultants for the Department of Trade and Industry, London.

Goyder, J., (1997). *Technology and Society: A Canadian Perspective*, Peterborough, Ontario, Broadview.

Granovetter, M. (1973). The Strength of Weak Ties, *American Journal of Sociology* 78(6), 1360-1380.

Gray, J., (1995). *Enlightenment's Wake*, London, Routledge.

Green, L., (1997). Issues Paper on Social Access to Learning Technologies, http://olt-bta.hrdc-drhc.gc.ca/online/sasumme.html.

Grossman, L.K., (1995). *The Electronic Republic: Reshaping Democracy in the Information Age*, New York, Viking.

Guilhon, B. et al. (1997). *Economie de la connaissance et organisations*. Paris, l'Harmattan.

Haas, P.M., (Ed.), (1992). *Knowledge, Power and International Policy Coordination*, Columbia, SC, University of South Carolina Press.

Habermas, J., (1974). *Legitimation Crisis*, Boston, Beacon Press.

Hagedoorn, J., (1993). "Understanding the Rationale of Strategic Technology Partnering: Interorganizational Modes of Cooperation and Sectoral Differences", *Strategic Management Journal*, Vol. 14, Pp. 371-85.

Halstead, T., (1999). "A Politics for Generation X", *The Atlantic Monthly*, Vol. 284, No. 2, August, Pp. 33-43.

Hamdani, D., (1997). "Business Demographics as Indicators of Innovation Activity", Statistics Canada, Science and Technology Redesign Project Working Paper No. ST-97-12.

Hamdani, D., (1998). "Business Demographics: Volatility and Change", *Services Indicators*, Statistics Canada Cat. No. 63-016-XPB, 2nd quarter, reprinted in *Canadian Economic Observer*, Statistics Canada, Catalogue No. 11-010, 11.3, March, 1998.

Hamdani, D., (1999). "Methodology of Innovation Surveys: Implications for Data Interpretation", Working paper, Science, Innovation and Electronic Information Division, Statistics Canada.

Handy, C., (1989). *The Age of Unreason*, Boston, Harvard Business School Press.

Harman, W.W., (1978). "Chronic Unemployment: An Emerging Problem of Postindustrial Society", *Futurist*, Vol. 12, Pp. 209-214.

Hart, A., (1996). "Howdy Partner", *Internet World*, Vol. 7, No. 11, http://www.iworld.com

Haughety, M. & Anderson, T., (1998). *Networked Learning: The Pedagogy of the Internet*, Montréal, Chenelière/McGraw-Hill.

Hayek, F.A. von, (1945). "The Use of Knowledge in Society", *American Economic Review*, 35, September, Pp. 519-530.

Health Canada, (1999). "Health Infoway", Final report of the Advisory Council on Health Infostructure, February.

Heisig, U. & Littek, W., (1995). "Trust as a Basis of Work Organisation", in Littek, W. & Charles, T. (Eds.), *The New Divison of Labour: Emerging Forms of Work Organisation in International Perspective*, Pp. 17-56, Berlin, Walter de Gruyter.

Hill, G.C., (1995). "The Myth of Multimedia", *The Wall Street Journal*, June 19, P. R6.

Hirschhorn, L. & Mokray, J., (1992). "Automation and Competency Requirements in Manufacturing: A Case Study", in Adler, P.S., *Technology and the Future of Work*, Pp. 15-45, Washington, DC, The Brookings Institution.

Hirst, P. & Zeitlin, J., (1991). "Flexible Specialization Versus Post-Fordism: Theory, Evidence and Policy Implications", *Economy and Society*, Vol. 20, Pp. 1-56.

Hollingsworth, R. (1993). Variation Among Nations in the Logic of Manufacturing Sectors and International Competitiveness. In D. Foray and C. Freeman (eds) *Technology and the Wealth of Nations*. London, Pinter, 301-321.

Howell, D.R. and Wieler, S.S., (1998). Skill-biased Demand Shifts and the Wage Collapse in the United States: a Critical Perspective. *Eastern Economic Journal* 24 (3), 343-366.

Howells, J., (1997). "Research and Technology Outsourcing", *Technology Analysis and Strategic Management*, Vol. 11, No. 1, Pp. 17-29.

Huber, G., (1991). "Organizational Learning: The Contributing Processes and the Literatures", *Organization Science*, Vol. 2.

Hudson, W.J., (1993). *Intellectual Capital: How to Build it, Enhance it, Use it*, New York, Wiley.

Huntington, S.P., (1996). *The Clash of Civilizations: The Making of the World Order*, New York, Simon and Schuster.

Hutchins, E. (1991). Organizing Work by Adaptation. *Organization Science* 2(1).

Huxley, T.H., 1880 (1963; 1937). "Science and Culture", in *Essays: English and American*, (Registered edition), New York, Collier.

Industry Canada, (1995). "Connection, Community, Content: The Challenge of the Information Highway", Final Report Phase I of the Information Highway Advisory Council, http://strategis.ic.gc.ca/SSG/ih01070e.html.

Industry Canada, (1997). "Preparing Canada for a Digital World. Final Report Phase II of the Information Highway Advisory Council", http://strategis.ic.gc.ca/ SSG/ih01650e.html.

Industry Canada, (1998). "Smart Communities", November.

Industry Canada, (1998). *The Canadian Electronic Commerce Strategy*, Ottawa, http:/e-com.ic.gc.ca.

Information Highway Advisory Council, (1995). "Connection, Community, Content: The Challenge of the Information Highway", final report of the Information Highway Advisory Council, Ottawa.

Information Highway Advisory Council, (1997). "Preparing Canada for a Digital World: Phase II, Conclusions and Recommendations", Information Highway Advisory Council, Ottawa, April.

Innis, H., (1951). *The Bias of Communication*, Toronto, University of Toronto Press.

Jacquemain, A., (Ed.), (forthcoming). *Intangibles and Competitiveness: a Challenge for Europe*, Aldershot, Edward Elgar.

Jacquemin, A. (1995). Capitalism, Competition, Cooperation. *De Economist* 143(1), 1-14.

Jaeger, C. & Ernste, H., (1989). "Ways Beyond Fordism?", in Ernste, H. & Jaeger, C. (eds.), *Information Society and Spatial Structure*. London, Belhaven Press.

Jaeger, C. (Ed.), *Information Society and Spatial Structure*, London, Belhaven Press.

Janz, N., Ebling, G., Gottschalk, S. & Niggemann, H., (1999). "Innovation Activity of the German Economy: Manufacturing and Mining Survey 1998", Mannheim, ZEW.

Janz, N., Ebling, G., Gottschalk, S. & Niggemann, H., (1999). "Services in the Future: Innovation Activity in the German Service Sector", Mannheim, ZEW.

Jhingran, A., (1998). "Data Mining and Electronic Commerce", Rutgers University Presentation, February.

Johnson, G.F., (1997). "Changes in Earnings Inequality: The Role of Demand Shifts", *Journal of Economic Perspectives*, Vol. 11, Pp. 41-54.

Jorgenson, D., (1997). "Computers and Productivity", paper presented to the Centre for the Study of Living Standards conference on "Service Sector Productivity and the Productivity Paradox", April 11-13, 1997, Ottawa, Canada, July-August.

Joshi, K. and Lauer, T.W., (1998). Impact of information technology on users' work environment: a case of computer aided design (CAD) system implementation. *Information and Management* 34 (6), 349-360.

Jovanovic, B. and Nyarko, Y., (1996). Learning by doing and the choice of technology. *Econometrica* 64 (6), 1299-1310.

Kahneman, D. and A. Tversky, (1979), Prospect Theory: an Analysis of Decision Under Risk, *Econometrica* 47 (2),263-291.

Kaiser, U., (1998). *Shifting Employment Patterns*, Mannheim, ZEW-Discussion Paper 98-26.

Kaplan, R.S. & Norton, D.P., (1996). *The Balanced Scorecard: Translating Strategy into Actions*, Boston, Harvard Business School Press.

Katz J. & Aspden P., (1998). "Internet Dropouts in the USA", *Telecommunications Policy*, Vol. 22, No. 4/5, July.

Kazancigil, A., (1998). "Governance and Science: Market-Like Models of Managing Society and Producing Knowledge", *International Social Science Journal*, Vol. 155, March, P. 69-79.

Keane, J., (1998). *Civil Society: Old Images, New Visions*, Cambridge, Polity Press.

Keasey, K., Thompson, S. & Wright, M., (1997). *Corporate Governance: Economic, Management and Financial Issues*, Oxford, Oxford University Press.

Kegan, R., (1994). *In Over Our Heads, The Mental Demands of Modern Life*, Cambridge, Massachusetts, Harvard University Press.

Kelly, K. 1998. *New Rules for the Economy*. New York, Penguin.

Kelly, T., (1997). "ITU to Release Report on Trade in Telecommunications", *International Telecommunications Union*, February 14, http: www.itu.ch.

Kennedy, D., (1999). "Background Noise? The Underlying Politics of Governance", *Harvard International Review*, Summer, Vol. XXI, No. 3, Pp. 52-57.

Keynes, J.M., (1919). *The Economic Consequences of the Peace*, London, Macmillan.

Killeck, T., (Ed.), (1995). *The Flexible Economy: Causes and Consequences of the Adaptability of National Economies*, London, Routledge.

Kingman, Sir John, (1998). "Science, Statistics and Public Policy", in Herzberg,
A. M. & Krupka, I., (Eds.), *Statistics, Science and Public Policy*, Pp. 7-12, Kingston, Ontario, Queen's University.

Klein, D.A., (1998). *The Strategic Management of Intellectual Capital*, Boston,

Kleinschmidt, M., Pekruhl, U., (1994). *Kooperative Arbeitstrukturen und Gruppenarbeit in Deutschland*, Institut Arbeit und Technik, Gelsenkirchen.

Kooiman, J. (Ed.), (1993). *Modern Governance: New Government-Society Relations*, London, Sage.

Krugman, P., (1994). *The Age of Diminished Expectations. U.S. Economic Policy in the 1990s*, (Revised and Updated Edition), Cambridge, Massachusetts, MIT Press.

Krugman, P.R. (1996). *The Self-Organizing Economy*. Oxford, Blackwell.

Krumenaker, L., (1996). "Surveyors of Cyberspace", *Internet World*, Vol. 7, No. 6, http://www.iworld.com.

Kuttner, R., (1996). *Everything for Sale: The Virtues and Limits of Markets*, New York, Knopf.

Lacroix, J-G & Tremblay, V., (1997). "The 'Information Society' and Cultural Industries Theory", *Current Sociology*, Vol. 45, No.4.

Lakoff, S.A., (1996). *Democracy: History, Theory and Practice*, Boulder, Westview.

Landauer, T.K., (1995). *The Trouble with Computers. Usefulness, Usability, and Productivity*, Cambridge, Massachusetts, MIT Press.

Lane, R.E., (1966). "The Decline of Politics and Ideology in a Knowledgeable Society", *American Sociological Review*, Vol. 31.

Laroche, M., Merette, M. & Ruggeri, G.C., (1999). "On the Concept and Dimensions of Human Capital in a Knowledge-Based Economy Context", *Canadian Public Policy/Analyse de Politiques*, Vol. XXV. No. 1, Pp. 87-100.

Le Bas, C. (1993). La firme et la nature de l'apprentissage. *Économies et Sociétés* Série Dynamique technologique et organisation, 1(5), 7-24.

Lee, F. & Has, H., (1996). "A Quantitative Assessment of High Knowledge Industries Versus Low Knowledge Industries", in Howitt, P. (Ed.), *The Implications of Knowledge-Based Growth for Microeconomic Policies*, University of Calgary Press.

Legler, H., Licht, G. & Spielkamp, A., (1999). "Germany's Technological Competitiveness, Report on Behalf of the German Ministry for Education and Research", Mannheim, Zentrum für Europäische Wirtschaftsforschung.

Leibenstein, H. (1976). *Beyond economic man*. Cambridge, Harvard University Press.

Leiss, W., (1990). *Under Technology's Thumb*, Montreal & Kingston, McGill-Queen's University Press.

Leonard, D. & Straus, S., (1997). "Putting your Company's Whole Brain to Work", *Harvard Business Review*, July-August.

Levitt, B. & March, J.G., (1988). "Organizational Learning", *Annual Review of Sociology*, Vol. 14.

Leydesdorff, L. and van den Besselaar, P. (Eds) 1994. *Evolutionary Economics and Chaos Theory* London, Pinter.

Licht, G. & Fier, A., (1999). "Management Training in Small and Medium Sized Company in Germany: A Case Study", Report on behalf the OECD's working party on SMEs, Mannheim.

Licht, G. & Janz, N. (Ed.), (1999). "Innovationsaktivität im Verarbeitenden Gewerbe und im Dienstleistungssektor", ZEW-Schriftenreihe, Bd. 38, Baden-Baden, Nomos.

Licht, G. & Moch, D., (1999). "Innovation and Information Technology in Services", *Canadian Journal of Economics*, Vol. 32, Pp. 363-383.

Licht, G., Hipp, C., Kukuk, M., Münt, G. & Janz, N., (1997). "Innovationen im Dienstleistungssektor. Empirischer Befund und Wirtschaftspolitische Konsequenzen", ZEW-Schriftenreihe, Bd. 22, Baden-Baden, Nomos.

Llerena, D. (1997). Coopérations cognitives et modèles mentaux collectifs : outils de création et de diffusion des connaissances, in B. Guilhon *et al.*, Eds, *Économie de la connaissance et organisations*. Paris L'Harmattan, 356-382.

Loeb, L., (1996). "The Stage is Set", *Internet World*, Vol. 7, No. 9, http://www.iworld.com.

Loveridge, D., Georghiou, L. & Nedeva, M., (1996). *Technology Foresight Programme: Delphi Survey*, Manchester, PREST, results available on web at htttp:\\www.foresight.gov.uk

Luhmann, N. (1993). *Risk: A Sociological Theory*, New York, Aldine de Gruyter.

Lund, R. & Gjerding, A. N., (1996). "The Flexible Company: Innovation, Work Organisation and Human Resource Management", mimeo; reported in more detail in Gjerding, A. N. (Ed.), (1997), in Danish, *Den fleksible virksomhed: Omstillingspres og fornyelse i dansk erhvervsliv*, The Danish Industry and Trade Development Council, Copenhagen.

Lundvall, B.-A. and Johnson, B., (1992). *The Learning Economy*, Paris, EAEPE conference on "Structural Change & the Regulation of Economic Systems".

Lundvall, B.-A. & Johnson, B., (1994). "The Learning Economy", *Journal of Industry Studies*, Vol. 1, No. 2.

Lundvall, B.-A., (1996). *The Social Dimensions of the Learning Economy*, Aalborg & Copenhagen, DRUID working paper 96-1.

Luttwak, E., (1998). *Turbo-Capitalism: Winners and Losers in the Global Economy*, London, Weidenfeld and Nicholson.

Machlup, F. (1980–84). *Knowledge: Its Creation, Distribution and Economic Significance*. Princeton, Princeton University Press. Vol. 1-3.

Malerba, F. (1992). Learning by Firms and Incremental Technical Change. *Economic Journal*, 102, 845-859.

Mallett, T., (1999), "Connecting to Small Business", presented at the Workshop on Statistical Challenges of Connectedness and Electronic Commerce, Ottawa, February 18-19.

Maloff, J., (1995). "The Virtual Corporation", *Internet World*, July 1995, http://www.iworld.com.

Manchester, P., (1997). "Complex Barriers to Electronic Commerce", *Financial Times*, April 2, P. 5.

Mansell, R. (Ed.), (1994). "Management of Information and Communications Technologies: Emerging Patterns of Control", Aslib, The Association for Information Management, London, U.K.

Mantei, M.M., Baecker, R.M., Sellen, A.J., Buxton, W.A.S., Milligan, T. & Wellman, B., (1994). "Experiences in the Use of a Media Space", mimeo, University of Toronto.

March, J.G. (1991). Exploration and Exploitation in Organizational Learning. *Organization Science* 2 (1), 71-87.

Marcuse, H., (1964). *One Dimensional Man*, Boston, Beacon Press.

Margherio, L., (1998). *The Emerging Digital Economy*, U.S. Department of Commerce, http://www.ecommerce.gov.

Marien, M. (1996). New Communications Technology: A Survey of Impacts and Issues. *Telecommunication Policy*, 20 (5), 375-387.

Marquardt, M. J., (1995). "Building a Global Learning Organization", *Industry and Higher Education*, August.

Marsick, V.J., (1998). "Transformative Learning From Experience in the Knowledge Era", *Daedalus*, Vol. 127, Pp. 119-136.

Martine, J., (1998). "Smart Managing: Lifelong Learning Spells Earnings", *Fortune*, July 6, Pp. 197-200.

Marx, K. & Engels, F., [1932] (1960). *Die deutsche Ideologie*, Berlin, Dietz.

Masie, E., (1999). *The 1999 Business of On-Line Learning Conference.* http://www.onlinelearning.org/busi2/ppt/Business%20of%20On-Line %20Learning%20'99/index.htm.

Massey, D., (1984). *Spatial Divisions of Labor: Social Structures and the Geography of Production*, London, Macmillan.

McCain, R.A. (1992). *A framework for cognitive economics.* Westport, Conn., Praeger.

McChesney, R.W., Wood, E.M. & Foster, J.B. (Eds.), (1998). *Capitalism and the Information Age: The Political Economy of the Global Information Revolution*, New York, Monthly Review Press.

McDaniel, S.A. & Tepperman, L., (1999). *Close Relations: Sociology of Families*, Toronto, Prentice Hall.

McKeever, J., (1996). "The Internet is the Second Revolution for the Industry", *Computer Reseller News*, Vol. 708, Nov. 4.

McKenzie, R.B. & Lee, D., (1991). *Quicksilver Capital*, New York, The Free Press.

McLuhan, M., (1962). *The Gutenberg Galaxy: The Making of Typographic Man*, Toronto, University of Toronto Press.

McMullen, K., (1998). "Skill and Employment Effects of Computer-Based Technology", Canadian Policy Research Networks Inc., CPRN Study No. W01, Ottawa.

Meisel, J., (1998). "Caesar and the Savants: Some Socio-Political Contexts of Science and Technology in Canada", in Herzberg, A. M. & Krupka, I. (Eds.), *Statistics, Science and Public Policy*, Pp. 153-177, Kingston, Ontario, Queen's University.

Merton, R.K., (1947). "The Machine, The Worker and the Engineer", *Science*, Vol. 105, Pp. 79-84.

Mesher, G., (1997). "The Internet in Asia", *Internet World*, Vol. 7, No. 12. http: www.iworld.com.

Miles, I., (1988). *Home Informatics: Information Technology and the Transformation of Everyday Life*, London, Pinter.

Miles, I., (1999). "Services and Foresight", *Service Industries Journal*, April 1999.

Miles, I. & Boden, M. (Eds.), (forthcoming 2000). *Knowledge, Innovation and Services*, London, Pinter.

Miles, I. & Matthews, M., (1992). "Information Technology and the Information Economy", in Robins, K. (Ed.), *Understanding Information Technology*, London.

Miles, I. & Tomlinson, M., (1999). "Intangible Assets and Service Sectors: The Challenges of Service Industries", paper presented at symposium: Intangible Assets And The Competitiveness Of The European Economy, 29th-30th April 1999, Université Catholique De Louvain (Ucl), Belgium; forthcoming in Jacquemain, A., (Ed.). *Intangibles and Competitiveness: a Challenge for Europe*, Aldershot, Edward Elgar.

Miles, I., Andersen, B., Boden, M. & Howells, J., (forthcoming). "Service Production and Property", *International Journal of Technology Management.*

Miles, I., Kastrinos, N., Flanagan, K., Bilderbeek, R., den Hertog, P. Huntink, W. & Bouman, M., (1995). *Knowledge-Intensive Business Services: - Users, Carriers and Sources of Innovation*, EC, Luxembourg, (DG13 SPRINT-EIMS).

Miles, I. with others, (1990). *Mapping and Measuring the Information Economy*, British Library.

Miller, R., (1996). *Measuring What People Know: Human Capital Accounting for the Knowledge Economy*, Paris, OECD.

Ministry of Labour, (1996). *Flexible Enterprise, Finnish Survey*, Helsinki.

Minkler, A.P. (1993). The Problem with Dispersed Knowledge: Firms in Theory and Practice. *Kyklos* 46(4), 569-587.

Moati, P. and Mouhoud, E.M. (1994). Information et organisation de la production : vers une division cognitive du travail. *Économie appliquée* 46(1), 47-73.

Mokhtarian, P.L., (1998). A synthetic approach to estimating the impact of telecommuting on travel. *Urban Studies* 35 (2), 215-241.

Moran, N., (1997). "New Payroll Service Via the Internet", *Financial Times*, February 5, P. 8.

Mulgan, G. (1997). *Connexity*. Cambridge, Harvard Business School Press.

Mullich J., (1997). "Winning Through Partnering", *Web Week*, Vol. 3I5, March 3. http://www.iworld.com.

Mumford, L., [1963] (1934). *Technics and Civilization*, New York, Harcourt, Brace and World.

Murphy, K., (1997). "Law Group Formulates Uniform Rules for On-Line Transactions", *Web Week*, Vol. 4I5, March 17, http://www.iworld.com.

Naim, G., (1997). "Turning the Net on its Head". *Financial Times*. March 5, P. 8.

Nairn, G., (1997). "A Crucial Year Ahead for the 'Net", *Financial Times*, January 8, P. 6.

Negroponte, N., (1995). *Being Digital*, New York, Vintage Books.

Nelson, R. & Phelps, E., (1966). "Investment in Humans, Technological Diffusion and Economic Growth", *American Economic Review*, Vol. 56, Pp. 69-75.

Nelson, R.R. & Wright, G., (1992). "The Rise and Fall of American Technological Leadership: The Postwar Era in Historical Perspective", *Journal of Economic Literature*, Vol. 30, 1931-1964.

Newing, R., (1997). "Electronic Cash Will Do Nicely, Sir", *Financial Times*, January 8, P. 7.

Newton, K., (1996). *The Human Factor in Firms' Performance: Management Strategies for Productivity and Competitiveness in the Knowledge-based Economy*, Occasional Paper No. 14, Industry Canada, Ottawa.

Newton, K., (1998). "The High Performance Workplace: HR-based management innovations in Canada", *International Journal of Technology Management*.

Newton, K. & Magun, S., (1998). *Organizational Learning and Intellectual Capital*, paper prepared for Industry Canada/CYRANO conference on "Doing Business in the KBE", Mont Tremblant, Quebec, September.

Nix, J. and Gabel, D., (1996). The Introduction of Automatic Switching into the Bell System: Market versus Institutional Influences. *Journal of Economic Issues* 30 (3), 737-753.

Nonaka, I., (1991). "The Knowledge-Creating Company" *Harvard Business Review*, November-December.

Nonaka, I. & Takeuchi, H., (1995). *The Knowledge Creating Company: How Japanese Companies Create the Dynamics of Innovation*, New York, Oxford University Press.

North, D.C. (1996). Economic performance through time. In:. Alston, L.J. Eggertsson, T. and North, D.C. (Eds) *Empirical studies in institutional change*. Cambridge, Cambridge University Press, 342-355.

O'Mahony, M., (1999). "Britain's Productivity Performance 1950-1996, An International Perspective", National Institute of Economic and Social Research, London.

O'Rourke, J. & Schachter, L., (1997). *The Janus Project - New Learning Technologies: Promises and Prospects for Women*, http://olt-bta.hrdc-drhc.gc.ca/publicat/januse.html

Organisation for Economic Cooperation and Development (OECD) (1992). *Technology and the Economy: The Key Relationships*, Paris, OECD.

OECD, (1993). *Proposed Standard Practice for Surveys of Research and Development - The FRASCATI Manual*, Paris, OECD.

OECD - ICCP, (1993). *Usage Indicators - A New Foundation for Information Technology Policies*, Paris, Organisation for Economic Cooperation and Development, Information Computer Communications Policy ICCP 31.

OECD, (1996). *Technology, Productivity and Job Creation*, Paris, OECD.

OECD. (1996). *Employment and Growth in the Knowledge-Based Economy*. Paris, OECD.

OECD, (1997). "Communications Outlook 1997", Vol. 1, Paris, OECD

OECD, (1997). "Global Information Infrastructures - Global Information Society (GII - GIS): Policy Requirements", Directorate for Science, Technology and Industry, Committee for Information, Computer and Communication Policy, Paris, OECD, February.

OECD, (1997). "The Role and Importance of Small and New Technology-based Firms", internal working paper, Paris, OECD.

OECD, (1997). *Guidelines for Collecting and Interpreting Innovation Survey Data – The OSLO Manual, Second Edition*, Paris, OECD.

OECD, (1997). *Implementing the OECD Jobs Study: Lessons from Member Countries' Experience*, Paris, OECD.

OECD, (1997). *Lifelong Learning to Maintain Employability*, OCDE/GD(97)162, Paris.

OECD, (1997). *Oslo Manual: Proposed Guidelines for Collecting and Interpreting Technological Innovation Data*, Paris, OECD/Eurostat.

OECD, (1998), *The Economic and Social Impact of Electronic Commerce*, Paris,

OECD, (1998). "Global Electronic Commerce: Realising the Potential", forward by the Right Honourable Jean Chretien, Prime Minister of Canada, October, OECD.

OECD, (1998). *Education at a Glance*, Paris, OECD.

OECD, (1998). *Meeting of the Industry Committee at Ministerial Level: Scoreboard of Indicators*, Paris, OECD.

OECD, (1998). *Pathways and Participation in Vocational and Technical Education and Training*, Paris, OECD.

OECD, (1998). *Technology, Productivity and Job Creation. Best policy practices*, Paris, OECD.

OECD, (1999). *Employment Outlook*, Paris, OECD.

OECD, www.oecd.org/subject/e.commerce.

Offe, C., (1982). *Disorganized Capitalism*, Cambridge MA., MIT Press.

Office of Learning Technologies, (1997). *Online Forum on Workplace Learning Centres*, Final Report, http://olt-bta.hrdc-drhc.gc.ca/online/summar.html, and Final Summary, http://olt-bta.hrdc-drhc.gc.ca/online/WLCFin.html.

Office of Learning Technologies, (1998). "Models of Community Learning Networks in Canada", http://olt-bta.hrdc-drhc.gc.ca/publicat/index.html.

Office of Learning Technologies, (1998). *Learning on the Information Highway: A Learner's Guide to the Technologies*, Montréal, Chenelière/McGraw-Hill.

Ogburn, W.F., (1973). "The Hypothesis of Cultural Lag", in Williams, W.W. (Ed.) *Sociological Theory: An Introduction*, Pp. 140-144, Chicago, Aldine.

Ontario Premier's Council, (1995). *Performance Measures in the New Economy*.

Pal, L.A., (1999). "Wired Governance: The Political Implications of the Information Revolution" in Boyce, R. (Ed.), *The Communications Revolutions at Work: The Social, Economic and Political Impacts of Technological Change*, Montreal-Kingston, McGill-Queen's University Press.

Paquet, G. (1989). Liberal Education as Synecdoche, in C. Andrew and S.B. Esbensen, Eds. *Who's Afraid of Liberal Education?* Ottawa, University of Ottawa Press, 1-20.

Paquet, G. (1994). From the Information Economy to Evolutionary Cognitive Economics, in R.E. Babe, Ed., *Information and Communication Economics*. Boston, Kluwer Academic Publishers, 34-40.

Paquet, G. (1998). Evolutionary Cognitive Economics . *Information Economics and Policy* 10(4), 343-357.

Paquet, G. (1999a). *Governance through Social Learning*. Ottawa, The University of Ottawa Press.

Paquet, G., (1999b). "Distributed Governance and Transversal Leadership", *Governance Through Social Learning*, Ottawa, University of Ottawa Press.

Paquet, G., (1999c). "The Strategic State", *Governance Through Social Learning*, Ottawa, University of Ottawa Press.

Patterson, R., (1997). "Review of Cutting Edge: Technology, Information, Capitalism and Social Revolution", edited by Davis, J., Hirschl, T. & Stack, M., *Contemporary Sociology*, Vol. 28, No. 2, Pp.185-186. New York, Verso, 1997.

Paye, J.-C., (1996). "Policies for a Knowledge-Based Economy", *OECD Observer*, July.

Petit, P. & Soete, L., (1997). "Is a Biased Technological Change Fuelling Dualism", paper presented to the Centre for the Study of Living Standards conference on "Service Sector Productivity and the Productivity Paradox", April 11-13, 1997, Ottawa, Canada.

Petrillo, J., (1995). "Keynote address to Internet@Telecom95", Internet @Telecom 95 October 8.

Pinsonneault, R.S., (1998). "Information Technology and the Nature of Managerial Work: From the Productivity Paradox to the Icarus Paradox?", *MIS Quarterly*, Vol. 22, Pp. 287-311.

Piore, M.J. & Sabel, C.F., (1984). *The Second Industrial Divide*, New York, Basic Books.

Porter, M., (1985). *Competitive Advantage*, New York, The Free Press.

Prusak, L., (1997). *Knowledge in Organizations*, Boston, Butterworth-Heinemann.

Pryor, F.L. & Schaffer, D.L., (1999). *Who's Not Working and Why. Employment, Cognitive Skills, Wages, and the Changing U.S. Labor Market*, Cambridge, Cambridge University Press.

Quinn, J.B., (1992). *Intelligent Enterprise: A Knowledge and Service Based Paradigm for Industry*, New York, Free Press.

Quinn, J.B., (1996). "The Productivity Paradox is False: Information Technology Improves Service Performance", *Advances in Services Marketing and Management*, Vol. 5, Pp. 71-84.

Ravetz, J.R., (1979). *Scientific Knowledge and its Social Problems*, Oxford, Oxford University Press.

Rhodes, R.A.W., (1996). "The New Governance: Governing Without Government", *Political Studies*, Vol. XLIV, Pp. 652-667.

Richards, G. & Goh, S.C., (1995). "Implementing Organizational Learning: Toward a Systematic Approach", *Optimum*, Autumn.

Richta, R., (1977). "The Scientific and Technological Revolution and the Prospects of Social Development", in Dahrendorf, R., *et. al.*, (Eds.), *Scientific Technological Revolution: Social Aspects*, London, Sage.

Ritter, J.B., (1992). "Defining International Electronic Commerce", *Northwestern Journal of International Law & Business*, Vol. 13, No. 1, Pp. 3-30.

Roberts, B., (1996). "HP's Net Strategy: Total Customer Support", *Web Week*, Vol. 2, No. I20, December 20, http://www.iworld.com.

Rohan, J., (1986). *Comparative and Multinational Management*, New York, John Wiley & Sons.

Rose, A., (1997). *Transfer of Funds for Research and Development in Canadian Industry 1993*, Ottawa, Statistics Canada Working Paper ST-97-05.

Rosenau, J.N., (1997). *Along the Domestic-Foreign Frontier: Exploring Governance in a Turbulent World*, Cambridge, Cambridge University Press.

Royal Bank, (1996). *10 Minute Guide to Canada's Transition to a Knowledge-Based Economy*, Customer Publication.

Ruigrok, E. & Van Tulder, R., (1995). *The Logic of International Restructuring*, London, Routledge.

Ryan, M.P., (1998). *Knowledge Diplomacy: Global Competition and the Politics of Intellectual Property*, Washington, Brookings Institution Press.

Sabel, C.F., (1991). "Moebius-Strip Organizations and Open Labor Markets: Some Consequences of the Reintegration of Conception and Execution in a Volatile Economy", in Bourdieu, P. & Coleman, J.S. (Eds.), *Social Theory for a Changing Society*, Pp. 23-54, Boulder, Westview Press.

Saint-Onge, H. (1996). Tacit Knowledge: The Key to the Strategic Alignment of Intellectual Capital. *Strategy and Leadership,* March-April, 10-14.

Saint-Onge, H., (1996). "The Learning Organization at CIBC...a framework for application", presentation at the "17th Annual National Business Conference", McMaster University, January 25.

Salomon, J.J., (in preparation). *The Technological Destiny*, London, Pinter.

Sanchez, R. & Aime, H., (1997). *Strategic Learning and Knowledge Management*, New York, Wiley.

Savage, C.M., (1996). *5th Generation Management: Co-creating Through Virtual Enterprising Dynamic Training and Knowledge Networking*, Toronto, Butterworth-Heinemann.

Scace, E., (1998). "International Development of UUNET Technologies", Rutgers University on Global Telecommunications, May 21.

Schelling, T. 1978. *Micromotives and macrobehavior*. New York, Norton.

Schelsky, H., (1954). "Zukunftsaspekte der industriellen Gesellschaft", *Merkur*, Vol. 8, Pp. 13-28.

Scholve, D., (1996). *Technology and Democracy*, New York, Prometheus.

Schon, D.A. 1983. *The Reflective Practitioner*. New York, Basic Books.

Schrödinger, E., (1952). "Are There Quantum Jumps?", *The British Journal for the Philosophy of Science*, Vol. 11.

Schumpeter, J.A., (1934). *The Theory of Economic Development*, New York, Harper & Row.

Segal, R., (1995). "Electronic Commerce Winners and Losers", *Planning Review*, Vol. 23, No. 6, P. 26, Nov/Dec.

Senge, P., (1990). *The Fifth Discipline: The Art and Practice of the Learning Organization*, New York, Doubleday.

Senker, J. (1995). Tacit Knowledge and Models of Innovation. *Industrial and Corporate Change*, 4(2), 425-447.

Shafe, L., (1997). "Technology Should Enrich Humanity, Not Diminish It" *Financial Times*, January 8, P. 7.

Shapiro, C. & Varian, H.R., (1998). *Information Rules: A Strategic Guide to the Network Economy*, Cambridge, MA., Harvard Business School Press.

Shillingford, J., (1997). "Why Intranets Are Catching On", *Financial Times*, April 2, P. 11.

Sichel, D.E., (1999). "Computers and Aggregate Economic Growth: An Update", *Business Economics*, Vol. 34, Pp. 18-24.

Silverstone, R. & Haddon, L., (1996). "Design and the Domestication of Information and Communication Technologies: Technical Change and Everyday Life", in Mansell, R. & Silverstone, R., *Communication by Design*, Oxford, Oxford University Press.

Silvestrou, R., Fitzgerald, L., Johnston, R. & Grant, C., (1992). "Toward a Classification of Service Processes", *International Journal of Service Industry Management*, Vol. 3, No. 3, Pp. 62-75.

Skandia, (1996). *Customer Value: Intellectual Capital*, supplement to Skandia's 1996 annual report.

Skyrme, D., (1998). *Measuring the Value of Knowledge: Metrics for the Knowledge-Based Business*, London, Business Intelligence Ltd. *Sociology of Capitalism*, Princeton, Princeton University Press.

Soete, L., (1997). *Some Thoughts on the 'Economics of Knowledge'*, paper prepared for the Advisory Committee on Science & Technology, Industry Canada, Ottawa.

Sombart, W., [1927] (2000). "The Economic Life of the Future", in Sombart, W., *Economic Life in the Modern Age*, edited and introduced by Stehr, N. & Grundmann, R., New Brunswick, Transaction Books.

Stahel, W. R., (1998). "From Products to Services: Selling Performance Instead of Goods", *The IPTS Report*, No. 27, September 1998, Pp. 35-42.

Statistics Canada, (1997). "Education in Canada", Catalogue No. 81-229-XIB.

Statistics Canada, (1997). "Household Facilities and Equipment, 1997", http://www.statcan.ca/english/Pgdb/People/Families/famil10b.htm.

Statistics Canada, (1998), *Statistics Canada Daily*, July 8.

Statistics Canada, (1998). "Population Estimates for 1996 and Projections for the Years 2001, 2006, 2011, and 2016", CANSIM, Matrices 6367 and 6900.

Statistics Canada, (1998). "Family Expenditure in Canada (FAMEX), 1996", Cat. No. 62-555-XPB, July.

Statistics Canada, (1999a). *Science and Technology Activities and Impacts: A Framework for a Statistical Information System*, Ottawa, Statistics Canada, Catalogue No. 88-522-XPB.

Statistics Canada, (1999b). *A Five-Year Strategic Plan for the Development of an Information System for Science and Technology*, Ottawa, Statistics Canada, Catalogue No. 88-523-XPB.

Stehr, N. & Ericson, R.V. (Eds.), (1992), *The Culture and Power of Knowledge*, New York, Walter de Gruyter.

Stehr, N., (1994). *Knowledge Societies*, London, Sage.

Stehr, N., (1999). "Deciphering Information Technologies: Modern Societies as Networks", *European Journal of Social Theory*

Stehr, N., (1999). "The Productivity Paradox: ICTs, Knowledge and the Labour Market", in de la Mothe, J. & Paquet, G. (Eds.), *Information, Innovation and Impact*, Kluwer, Boston, Massachusetts.

Stewart, T.A., (1997a). "Tools for Measuring and Managing Intellectual Property", in Stewart, T.A., *Intellectual Capital: The New Wealth of Organizations*, Pp. 222-246, New York, Doubleday.

Stewart, T.A., (1997b). *The New Organizational Wealth: Managing and Measuring Knowledge-Based Assets*, San Francisco, Berrett-Kohler.

Stivers, B., et. al., (1996). *The Management of Intangible Assets: A Study of US and Canadian Firms*, prepared for Canadian Institute of Chartered Accountants.

Stoffman, D., (1997). *Masters of Change*, Toronto, McGraw-Hill Ryerson.

Stoll, C., (1995). *Silicon Snake Oil*, Anchor Books, New York.

Storper, M. (1996). Institutions of the Knowledge-Based Economy, in *Employment and Growth in the Knowledge-Based Economy*. Paris, OECD, 255-283.

Strange, S., (1996). *The Retreat of the State: The Diffusion of Power in the World Economy*, Cambridge, Cambridge University Press.

Sveiby, K.E., (1997). *The New Organizational Wealth*, San Francisco, Berrett-Koehler.

Swedberg, R. (Ed.), (1991). *Joseph A. Schumpeter: The Economics and*

Taylor, P., (1997). "Confident of a More Colourful Future", *Financial Times*, March 5, P. 2.

Tenner, E., (1996). *Why Things Bite Back: Technology and the Revenge of Unintended Consequences*, New York, Knopf.

The Economist, 14 June, www.economist.com/editorial/freeforall/current/ wb8852.html.

The Economist, (1999). "The End of Privacy", *The Economist*, May 1-7, Pp. 21-

The Economist, (1999). "Measuring Intangible Assets: A Price on the Priceless", June 14.

The Economist, (1999). "How Real is the New Economy?", *The Economist*, July 24-30, Pp. 21-24, P. 67.

The Economist, (1999). "Politics in Brief, Is There a Crisis?", *The Economist*, July 17-23, P. 49-50.

Thurbin, P., (1995). *Leveraging Knowledge*, London, Pitman Publishing.

Thurow, L.C., (1999). *Building Wealth: The New Rules for Individuals, Companies and Nations in a Knowledge-Based Economy*, New York, Harper Collins.

Tidd, J., Bessant J. & Pavitt, K., (1997). *Managing Innovation: Integrating Technological, Market and Organisational Change*, Chichester, United Kingdom, John Wiley & Sons.

Tilly, Chris & Tilly, Charles, (1998). *Work Under Capitalism*, Boulder, Colorado, Westview Press.

Tobin, D., (1996). *Transformation Learning*, New York, Wiley.

Tobin, D., (1998). *The Knowledge-Enabled Organization*, New York, Amacom.

Tricker, R.I., (1984). *International Corporate Governance*, Englewood Cliffs, Prentice Hall.

Ubios, J., (1997). The Eye of the Storm, *Internet World*, Vol. 8, No. 1, http://www.iworld.com

Uchupalanan, K., (1998). *Dynamics of Competitive Strategy and IT based Product-Process Innovation in Financial Services: The Development of Electronic Banking Services in Thailand*, DPhil thesis, University of Sussex, Falmer, Brighton.

United States, Department of Commerce, (1991). *U.S. Industrial Outlook*, Washington, DC.

United States, Department of Commerce, (1995). *Falling Through the Net: A Survey of the 'Have Nots' in Rural and Urban America*, Washington, DC.

United States, Department of Commerce, (1999). *Telecommunications and Information Infrastructure Assistance Program*, February

United States, Telecommunications and Information Administration, (1998). *Falling Through the Net II: New data on the Digital Divide*, National, July.

Valéry, N. (1999) "Innovation in Industry" *The Economist* (Survey published on February 20th)

Van Langenhove, L., (1999). "Rethinking the Social Sciences? A Point of View", in *The Social Sciences at a Turning Point?*, Pp. 43-51, Paris, OECD.

Vickery, G. & Murphy, M., (1998). *An International Study of Selected Business Services*, Paris, OECD.

Vickery, G. & Wurtzburg, G., (1992). "Intangible Investment: Missing Pieces in the Productivity Puzzle", *OECD Economic Observer*, Paris.

Vickery, G. & Wurzburg, G., (1998). "The Challenge of Measuring and Evaluating Organisational Change in Enterprises", in *Measuring Intangible Investment*, OECD, Paris.

Wagner, E.D., (1999). "Beyond Distance Education: Distributed Learning Systems", in Stolovitch, H. & Keeps, E. (Eds.), *Handbook of Human Performance Technology: Improving Individual and Organizational Performance*, San Francisco, Jossey-Bass, Pp. 626-650,

Waligorski, C.P., (1997). *Liberal Economics and Democracy: Keynes, Galbriath, Thurow, and Reich*, Lawrence, University Press of Kansas.

Wallerstein, M.B., Mogee, M.E. & Schoen, R.A. (Eds.), (1993). *Global Dimensions of Intellectual Property Rights in Science and Technology*, Washington, National Academy Press.

Waters, R., (1997). "New Sales Vehicle Revs Up", *Financial Times*, February 12, P. 21.

Weber, M., [1913] (1981). "Some Categories of Interpretive Sociology", *The Sociological Quarterly*, Vol. 22, Pp. 151-180.

Weisberg, J., (1996). *In Defence of Government*, New York, Scribner.

Weiser, M. (1991). "The Computer for the Twenty-First Century", *Scientific American,* September 1991, Pp. 94-100.

Wellman, B. & Gulia, M., (1998). "Virtual Communities as Communities", in Smith, M. & Kollock, P. (Eds.), *Communities in Cyberspace*, London, Routledge.

Wetzel, J.R., (1995). "Labor Force, Unemployment, and Earnings", in Farley, R. (ed.), *State of the Union. America in the 1990s.* Volume One, Economic Trends, New York, Russell Sage Foundation.

Wheelwright, G., (1997). "Experiments in Retailing", *Financial Times*, March 5, P. 8.

Wick, C. & Leon, L.S., (1993). *The Learning Edge*, McGraw Hill. New York

Wikstrom, S. and Normann, R. 1994. *Knowledge and Value — A New Perspective on Corporate Transformation*. London, Routledge.

Williams, R, and Edge, D. (1996). The Social Shaping of Technology. *Research Policy*, 25, 865-899.

Williamson, O., (1975). *Markets and Hierarchies: Analysis and Antitrust Implications*, New York, The Free Press.

Winer, L.R., Rushby, N. & Vázquez-Abad, J., (1999). "Emerging Trends in Instructional Interventions", in Stolovitch, H. & Keeps, E. (Eds.) *Handbook of Human Performance Technology: Improving Individual and Organizational Performance Worldwide*, San Francisco, Jossey-Bass.

Wolff, N.E. & Baumol, W.J., (1989). "Sources of Postwar Growth of Information Activity in the United States", in Osberg, L., Wolff, N.E. & Baumol, W.J. (Eds.), *The Information Economy: The Implications of Unbalanced Growth*, Institute for Research on Public Policy.

Wymbs, C., (1996). "Spatial Determinants of Sector Investment in an Emerging Information Economy", Rutgers University CIBEAR Working Paper, Newark, New Jersey.

Wymbs, C., (1997). "Does Cross Cultural Contact Associated with International Business Affect Managers' Values and Negotiating Outcomes? A Sino-United States Comparison", *American Society of Business and Behavioral Sciences Proceedings*, Las Vegas, February.

Wymbs, C., (1997). "Internet: Its Effect on International Trade, Foreign Direct Investment and Strategic Alliances", presented at the "Global Networking 1997" conference, Calgary, Canada, June 15-18.

Yoder, S. K., (1995). "Back in the Running", *The Wall Street Journal*, June 19, P. r22.

Young, A., (1996). *Measuring R&D in the Services*, STI Working Papers, Paris, OECD.

Zeller, S., (1997). "The Squabble Over Internet Taxes", *National Journal*, Vol. 29, No. 50, Pp. 2520-2521, December 13.

Ziman, J. (1991) "A Neural Net Model of Innovation" *Science and Public Policy*, 18, 1, 65-75.

Index

Economics of Science, Technology and Innovation

1. A. Phillips, A.P. Phillips and T.R. Phillips:
   Biz Jets. Technology and Market Structure in the Corporate Jet Aircraft Industry. 1994 — ISBN 0-7923-2660-1
2. M.P. Feldman:
   The Geography of Innovation. 1994 — ISBN 0-7923-2698-9
3. C. Antonelli:
   The Economics of Localized Technological Change and Industrial Dynamics. 1995 — ISBN 0-7923-2910-4
4. G. Becher and S. Kuhlmann (eds.):
   Evaluation of Technology Policy Programmes in Germany. 1995 — ISBN 0-7923-3115-X
5. B. Carlsson (ed.): *Technological Systems and Economic Performance: The Case of Factory Automation.* 1995 — ISBN 0-7923-3512-0
6. G.E. Flueckiger: *Control, Information, and Technological Change.* 1995 — ISBN 0-7923-3667-4
7. M. Teubal, D. Foray, M. Justman and E. Zuscovitch (eds.):
   Technological Infrastructure Policy. An International Perspective. 1996 — ISBN 0-7923-3835-9
8. G. Eliasson:
   Firm Objectives, Controls and Organization. The Use of Information and the Transfer of Knowledge within the Firm. 1996 — ISBN 0-7923-3870-7
9. X. Vence-Deza and J.S. Metcalfe (eds.):
   Wealth from Diversity. Innovation, Structural Change and Finance for Regional Development in Europe. 1996 — ISBN 0-7923-4115-5
10. B. Carlsson (ed.):
    Technological Systems and Industrial Dynamics. 1997 — ISBN 0-7923-9940-4
11. N.S. Vonortas:
    Cooperation in Research and Development. 1997 — ISBN 0-7923-8042-8
12. P. Braunerhjelm and K. Ekholm (eds.):
    The Geography of Multinational Firms. 1998 — ISBN 0-7923-8133-5
13. A. Varga:
    University Research and Regional Innovation: A Spatial Econometric Analysis of Academic Technology Transfers. 1998 — ISBN 0-7923-8248-X
14. J. de la Mothe and G. Paquet (eds.):
    Local and Regional Systems of Innovation — ISBN 0-7923-8287-0
15. D. Gerbarg (ed.):
    The Economics, Technology and Content of Digital T V — ISBN 0-7923-8325-7
16. C. Edquist, L. Hommen and L. Tsipouri
    Public Technology Procurement and Innovation — ISBN 0-7923-8685-X
17. J. de la Mothe and G. Paquet (eds.):
    Information, Innovation and Impacts — ISBN 0-7923-8692-2

KLUWER ACADEMIC PUBLISHERS — BOSTON / DORDRECHT / LONDON